Notes and Problems in Statistics

Also available from ST(P):

Crawshaw and Chambers	A CONCISE COURSE IN A-LEVEL STATISTICS
Francis	ADVANCED LEVEL STATISTICS
Greer	A FIRST COURSE IN STATISTICS
Greer	REVISION PRACTICE IN STATISTICS
Greer	STATISTICS FOR ENGINEERS
Montagnon	FOUNDATIONS OF STATISTICS
White et al.	TABLES FOR STATISTICIANS

Notes and Problems in Statistics

Richard Thomas AFIMA

Senior Lecturer in Statistics
Southampton Institute of Higher Education

Stanley Thornes (Publishers) Ltd

© Richard Thomas 1984

All rights reserved. No part of this publication may be reproduced, stored in a retrieval system or transmitted in any form or by any means, electronic, mechanical, recording or otherwise, without the prior written consent of the copyright holders. Applications for such permission should be addressed to the publishers: Stanley Thornes (Publishers) Ltd, Old Station Drive, Leckhampton, CHELTENHAM GL53 0DN, England.

First published in 1984 by
Stanley Thornes (Publishers) Ltd
Old Station Drive
Leckhampton
CHELTENHAM   GL53 0DN

British Library Cataloguing in Publication Data

Thomas, R
   Notes and problems in statistics
   1. Statistical mathematics
   I. Title
   519.5   QA276

ISBN 0-85950-135-3

Typeset by Tech-Set, Gateshead, Tyne and Wear.
Printed and bound in Great Britain at The Pitman Press, Bath.

CONTENTS

PREFACE	vii
ACKNOWLEDGEMENTS	viii
Chapter 1 **FREQUENCY DISTRIBUTIONS**	**1**
1.1 Construction of frequency distributions	1
1.2 Histograms	2
1.3 Frequency polygons	5
1.4 Cumulative frequency curves (ogives)	6
1.5 Further diagrams	8
Exercises on Sections 1.1 to 1.5	10
1.6 Arithmetic mean	12
1.7 Mode	13
1.8 Median	15
Exercises on Sections 1.6 to 1.8	17
1.9 Range	17
1.10 Inter-quartile range (IQR)	18
1.11 Mean deviation	20
1.12 Standard deviation	22
1.13 Coding	23
Exercises on Sections 1.9 to 1.13	24
Past questions 1	25
Chapter 2 **PROBABILITY**	**33**
2.1 Basic probability	33
2.2 Combination of events	33
Exercises on Sections 2.1 and 2.2	35
2.3 Combinations	36
2.4 Binomial distribution	37
2.5 Poisson distribution	39
Exercises on Sections 2.3 to 2.5	42
2.6 Normal distribution	42
Exercises on Section 2.6	47
Past questions 2	48
Chapter 3 **SAMPLING DISTRIBUTIONS**	**53**
3.1 Distribution of sample means	53
3.2 Significance of a sample mean	54
Exercises on Sections 3.1 and 3.2	56
3.3 The t distribution	57
3.4 Difference between two sample means	59
3.5 The paired t-test	62
Exercises on Sections 3.3 to 3.5	64
3.6 Distribution of proportions	65
Exercises on Section 3.6	68
Past questions 3	69

Chapter 4	THE χ^2 DISTRIBUTION	**74**
	4.1 Goodness of fit	74
	Exercises on Section 4.1	78
	4.2 Contingency tables	79
	4.3 Yates' correction	83
	Exercises on Sections 4.2 and 4.3	84
	Past questions 4	86
Chapter 5	CORRELATION AND REGRESSION	**91**
	5.1 Scatter diagrams	91
	5.2 A correlation coefficient	92
	5.3 Rank correlation coefficient	95
	Exercises on Sections 5.1 to 5.3	96
	5.4 Significance of the correlation coefficient	97
	5.5 Regression	99
	Exercises on Sections 5.4 and 5.5	103
	Past questions 5	104
Chapter 6	FORECASTING	**111**
	6.1 Trend	111
	6.2 Moving averages	112
	6.3 Seasonal variations	116
	Exercises on Sections 6.1 to 6.3	121
	6.4 The multiplicative model	122
	6.5 Exponential smoothing	127
	Exercises on Sections 6.4 and 6.5	129
	Past questions 6	130
Chapter 7	INDEX NUMBERS	**136**
	7.1 Price relatives	136
	7.2 Price aggregates	139
	7.3 Quantity/volume indices	143
	Exercises on Sections 7.1 to 7.3	145
	Past questions 7	148
ANSWERS		**154**
SUMMARY OF FORMULAE		**163**
INDEX		**167**

PREFACE

I believe that this book will aid students and lecturers involved in a wide variety of statistics courses related to business and management studies. The book contains numerous *graded* examples and worked solutions in addition to an abundance of exercises. It contains virtually no theory other than the list of formulae quoted at the beginning of each section. I expect the book to be particularly useful for revision purposes.

Primarily, the book is intended for 'non-mathematical' students. Consequently, I have avoided the use of formulae whenever possible. For example, in Chapter 1, although formulae exist for estimation values such as the mode, median, etc., I have only outlined the graphical methods of estimation. In Chapter 2, I have kept the basic probability examples as brief as possible and only covered sufficient ground for an introduction to the more important probability distributions. Chapter 3 is the most mathematical (and difficult) section of the book. Many students may decide to avoid it (!) and this would not drastically affect their comprehension of the remaining chapters. Chapters 4 to 7 are independent of each other and could be studied in any order. The significance of a correlation coefficient is examined in Chapter 5 using the t-test. However, I would expect many students to cover correlation without reference to significance. In such a case the student could merely ignore Section 5.4 without impairing his/her understanding of this important area.

I would recommend the use of *Tables for Statisticians* by White, Yeats and Skipworth. I have used these tables in the worked solutions throughout this text.

Finally, I must emphasise that all answers and worked solutions in this book are my own, and are not the responsibility of the various examining bodies involved. Many results in this book, particularly those obtained by graphical methods, are only estimates. Consequently, the reader should not be concerned if our answers do not agree exactly! All data contained in the problems (other than those from past examinations) given in this text are figments of my imagination and should not be taken too seriously!

RICHARD THOMAS

ACKNOWLEDGEMENTS

I would like to thank the following examining bodies for their permission to use the wide range of past examination questions in this text:

Association of Certified Accountants (ACA)
Chartered Institute of Public Finance and Accountancy (CIPFA)
Chartered Institute of Transport (CIT)
Chartered Insurance Institute (CII)
Institute of Cost and Management Accountants (ICMA)
Institute of Industrial Managers (IIM)
Institute of Management Services (IMS)
Institute of Marketing (IM)
Institute of Personnel Management (IPM)
Institute of Training and Development (ITD)
Rating and Valuation Association (RVA)
Royal Institution of Chartered Surveyors (RICS)

Note: Any past questions in this book that are not shown to have been obtained from one of the above examining bodies are internal College examination questions.

Furthermore, I would like to thank my colleagues in the Mathematics Department at Southampton College of Higher Education, particularly Malcolm and Barbara for their helpful comments, and Sheelagh for her patience and superb typing.

Finally, special thanks to my wife, Denise, and sons, Chris and Philip, for tolerating my antisocial behaviour during the writing of this book!

1 FREQUENCY DISTRIBUTIONS

1.1 CONSTRUCTION OF FREQUENCY DISTRIBUTIONS

EXAMPLE 1.1.1 The data below give the scores obtained in an aptitude test by a group of 40 applicants for a particular post in a company.

6	7	7	8	9	7	8	6	7	9
10	7	10	4	6	7	6	8	7	6
10	6	7	9	7	10	5	9	7	6
7	6	8	7	6	8	7	6	7	8

Construct a frequency distribution from this information.

SOLUTION Constructing a tally chart, we have:

Scores	Tally	Frequency
4	|	1
5	|	1
6	⊬⊬⊬ ⊬⊬⊬	10
7	⊬⊬⊬ ⊬⊬⊬ ||||	14
8	⊬⊬⊬ |	6
9	||||	4
10	||||	4
		Total = 40

We have the frequency distribution shown below.

Score, x	4	5	6	7	8	9	10
Frequency, f	1	1	10	14	6	4	4

EXAMPLE 1.1.2 Construct a grouped frequency distribution of the table of weekly wages earned by a sample of 50 employees shown below. (The figures are given in pounds (£).)

117	92	81	100	95	110	91	104	89	108
106	99	109	92	105	70	114	83	123	93
126	76	116	96	63	119	84	115	78	116
94	105	96	91	112	94	101	97	108	86
124	90	107	81	102	88	121	73	127	98

SOLUTION [*Note*: To help decide on class intervals, it is usual to construct a distribution with approximately 8-10 classes. However, this depends on the type and quantity of the original data, and in practice any number of classes may be desirable.]

The tally chart is shown below.

Wages (£)	Tally	Frequency
60–	I	1
70–	IIII	4
80–	⊥⊦⊦⊦ II	7
90–	⊥⊦⊦⊦ ⊥⊦⊦⊦ IIII	14
100–	⊥⊦⊦⊦ ⊥⊦⊦⊦ I	11
110–	⊥⊦⊦⊦ III	8
120–	⊥⊦⊦⊦	5
		Total = 50

The frequency distribution is:

Weekly wage, x (£)	60–	70–	80–	90–	100–	110–	120–
No. of employees, f	1	4	7	14	11	8	5

1.2 HISTOGRAMS

The histogram is the most common method of illustrating a frequency distribution. Each frequency in the distribution is represented by a block.

EXAMPLE 1.2.1 A survey of 50 retail outlets in the Southampton area gave the following distribution of apple prices.

Price (pence/lb)	28	29	30	31	32	33	34
Number of stores	1	1	5	21	13	6	3

Construct a histogram of these data.

SOLUTION

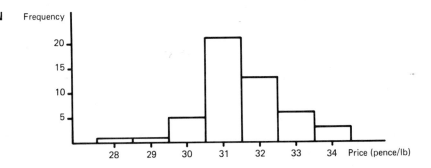

EXAMPLE 1.2.2 The results of a survey into the time taken to complete a given task by a group of 60 employees are given in the table below. Draw a histogram of the distribution obtained.

Time (minutes)	30–	35–	40–	45–	50–	55–	60–
No. of employees	2	17	18	13	6	1	3

SOLUTION

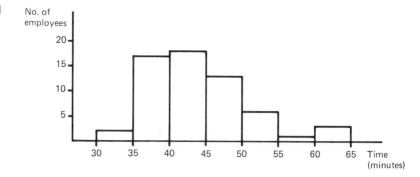

EXAMPLE 1.2.3 From the sales ledgers of a small company, the age of a sample of 100 debts are shown in the distribution below. Construct a histogram of this distribution.

Age of debt (days)	1–10	11–20	21–30	31–40	41–50	51–60
No. of accounts	22	33	25	14	4	2

SOLUTION [*Note*: There are no gaps between successive blocks on a histogram. Thus the blocks representing the classes 1–10 and 11–20 will be drawn to meet at $10\frac{1}{2}$ on the horizontal scale, etc.]

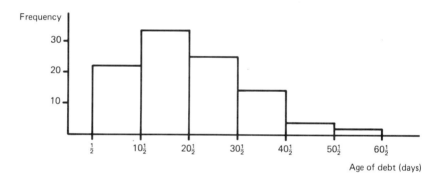

EXAMPLE 1.2.4 The age distribution of a random sample of 2000 people in Southampton is shown below. Construct a histogram from this table.

Age (years)	under 2	2–	5–	10–	20–	30–	40–	60–	80–
No. of people	96	105	169	327	309	276	409	260	49

SOLUTION [*Note*: The *area* of each block is proportional to the frequency that it represents. Consequently, in a distribution with unequal class intervals, the frequencies must be adjusted in order to obtain the heights of individual blocks.]

Frequencies are adjusted in the following table.

x	0–	2–	5–	10–	20–	30–	40–	60–	80–
f	96	105	169	327	309	276	409	260	49
Class width	2	3	5	10	10	10	20	20	20
Frequency density ($= f$/class width)	48	35	33.8	32.7	30.9	27.6	20.45	13	2.45

The histogram is shown below.

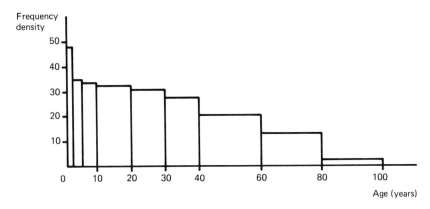

EXAMPLE 1.2.5 The table below gives the distribution of the length of spells of sickness absence for employees in a company over the past five years.

Duration of spell (days)	less than 3	3-5	6-8	9-14	15-20	21-27	28-50	over 50
No. of employees	342	267	291	320	264	140	230	165

Draw a histogram of the above data.

SOLUTION The frequency distribution is:

x	1-2	3-5	6-8	9-14	15-20	21-27	28-50	51-100
f	342	267	291	320	264	140	230	165

Closing the gaps, and adjusting the frequencies, we have:

x	$\frac{1}{2}$-$2\frac{1}{2}$	$2\frac{1}{2}$-$5\frac{1}{2}$	$5\frac{1}{2}$-$8\frac{1}{2}$	$8\frac{1}{2}$-$14\frac{1}{2}$	$14\frac{1}{2}$-$20\frac{1}{2}$	$20\frac{1}{2}$-$27\frac{1}{2}$	$27\frac{1}{2}$-$50\frac{1}{2}$	$50\frac{1}{2}$-$100\frac{1}{2}$
f	342	267	291	320	264	140	230	165
Class width	2	3	3	6	6	7	23	50
Frequency density	171	89	97	53.33	44	20	10	3.3

The histogram is shown below.

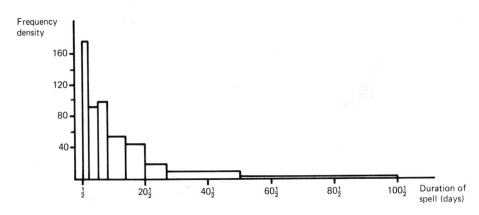

1.3 FREQUENCY POLYGONS

Frequency polygons can be used to illustrate frequency distributions. They are particularly useful when comparing two or more distributions.

EXAMPLE 1.3.1. The number of employees arriving late for work in a department over the past 180 days are given in the table below.

No. of employees late	0	1	2	3	4	5
No. of days	36	43	48	27	16	10

Draw a frequency polygon of these data.

SOLUTION

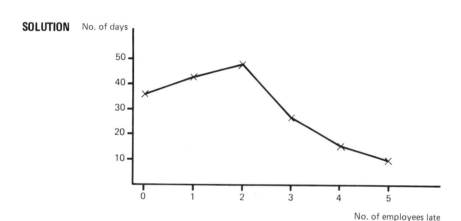

EXAMPLE 1.3.2 By drawing a suitable diagram, compare the distributions of wages for a sample of 100 employees in two companies.

Weekly wage (£)	60–	80–	100–	120–	140–
No. of employees:					
Company A	5	22	50	19	4
Company B	10	17	41	21	11

SOLUTION [*Note*: Points are plotted in the centre of each class interval.]

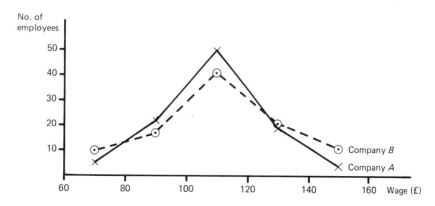

EXAMPLE 1.3.3 The following table gives the distribution of balances of a sample of 120 bank accounts in credit.

Balance (£)	0–	20–	50–	100–	200–	400–	1000–
No. of accounts	6	4	12	38	24	8	8

Draw a frequency polygon of this information.

SOLUTION [*Note*: The frequencies are adjusted in the same way as for a histogram.]

x	0–	20–	50–	100–	200–	400–	1000–
f	6	4	12	38	24	8	8
Class width	20	30	50	100	200	600	1000
Frequency density	0.3	0.133	0.24	0.38	0.12	0.013	0.008

The frequency polygon is shown below.

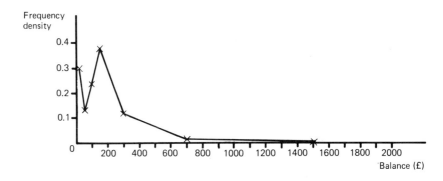

1.4 CUMULATIVE FREQUENCY CURVES (OGIVES)

Ogives are used to estimate a variety of information related to a frequency distribution.

EXAMPLE 1.4.1. Draw an ogive of the distribution of bonus payments made to 150 employees in a company shown below.

Monthly bonus (£)	0–	10–	20–	30–	40–	50–
No. of employees	28	43	38	29	7	5

SOLUTION Firstly, construct a cumulative frequency table, and then draw the ogive.

Bonus (£)	Cumulative frequency
0	0
10	28
20	71
30	109
40	138
50	145
60	150

FREQUENCY DISTRIBUTIONS 7

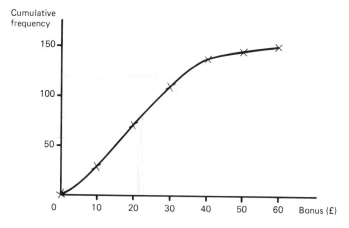

EXAMPLE 1.4.2 Draw an ogive of the distribution of wages of 250 employees given below, and hence estimate the number of employees earning (a) less than £70 per week and (b) more than £115 per week.

Weekly wage (£)	40–	50–	60–	80–	100–	120–	150–200
No. of employees	3	8	27	75	79	44	14

SOLUTION

Wage (£)	Cumulative frequency
40	0
50	3
60	11
80	38
100	113
120	192
150	236
200	250

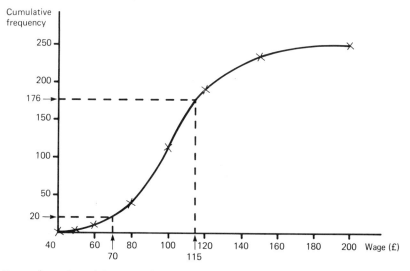

From the ogive, (a) 20 employees earn less than £70 and (b) 176 employees earn less than £115, so 74 employees earn more than £115.

1.5 FURTHER DIAGRAMS

EXAMPLE 1.5.1. Draw a *pie chart* to illustrate the expenditure of a large company on a number of advertising methods.

Method of Advertising	TV	Radio	Newspapers	Competitions	Others
Expenditure during 1982 (× £1000)	100	20	40	15	5

SOLUTION Total expenditure = £180 000. In a pie chart, £180 000 = 360°. The individual methods of advertising are represented by:

TV $\quad £100\,000 = 360° \times \dfrac{100\,000}{180\,000} = 200°$

Radio $\quad £20\,000 = 360° \times \dfrac{20\,000}{180\,000} = 40°$

Newspapers $\quad £40\,000 = 360° \times \dfrac{40\,000}{180\,000} = 80°$

Competitions $\quad £15\,000 = 360° \times \dfrac{15\,000}{180\,000} = 30°$

Others $\quad £5000 = 360° \times \dfrac{5000}{180\,000} = 10°$

The pie chart is shown below.

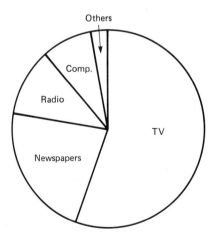

EXAMPLE 1.5.2 Use a *bar chart* to illustrate the number of workers employed in four factories as tabulated below.

Factory	A	B	C	D
No. of employees	120	300	250	150

FREQUENCY DISTRIBUTIONS 9

SOLUTION

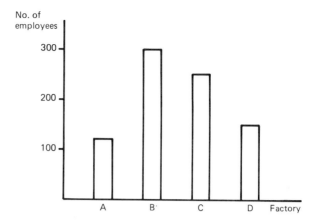

EXAMPLE 1.5.3 Draw a *component bar chart* of the data given below, for four factories, A, B, C and D.

	No. of employees			
	A	B	C	D
Unskilled	20	30	40	30
Semi-skilled	40	100	90	100
Skilled	60	170	120	20

SOLUTION

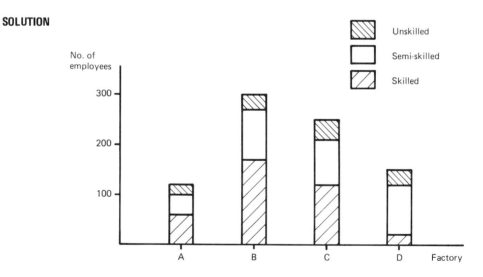

EXAMPLE 1.5.4 Draw a *multiple bar chart* to illustrate the performances of two companies over a five-year period.

	Output (× £1000)				
	1979	1980	1981	1982	1983
Company X	300	280	265	230	250
Company Y	190	240	270	260	240

SOLUTION

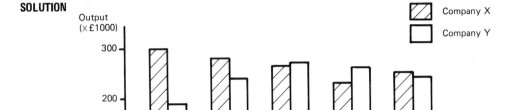

EXERCISES ON SECTIONS 1.1 TO 1.5

1. Draw a histogram and an ogive for both of the following grouped frequency distributions.

   (a)

x	0–	10–	20–	30–	40–
f	2	7	14	6	3

   (b)

x	0–	4–	8–	12–	20–	30–	40–	50–
f	3	8	11	16	15	12	4	2

2. The daily production figures for a particular manufactured item over the last 100 working days are shown in the following table.

No. of units produced	800–	850–	900–	950–	1000–	1050–	1100–
No. of days	3	4	15	25	43	6	4

   By drawing a suitable diagram, compare these figures with the distribution of production for the preceding 100 days shown below.

No. of units produced	850–	900–	950–	1000–	1050–	1100–
No. of days	6	17	29	39	6	3

   Using the combined figures for the last 200 days, estimate by means of an ogive the number of days where the production was (a) under 970 units and (b) over 1040 units.

3. A large company has two factories, A and B (managed independently). The Personnel Department has been asked to consider the problem of lateness in both factories by the unskilled workforce.

   The morning shift in both factories commences at 7.30 a.m. The workers clock in and are considered to be late if they clock in after 7.33 a.m. The data below give the lateness times, taken on one particular day in both factories.

	No. of minutes late (after 7.33 a.m.)								
Factory A	11	6	1	16	2	34	2	26	14
	20	5	31	7	21	1	7	6	49
	39	2	11	6	17	3	44	1	17
	3	20	1	25	5	18	2	29	8
	22	6	16	5	7	1	8	51	4
Factory B	24	4	49	8	18	12	7	20	1	23	9	38	5
	18	29	1	27	6	15	21	4	18	11	30	6	14
	56	6	20	2	16	9	42	12	27	5	17	14	21
	10	29	16	40	21	26	2	31	16	7	44	4	51
	19	3	3	12	7	11	18	1	23	6	33	17	9

Construct grouped frequency distributions for the lateness data in each factory and compare the distributions using a suitable diagram.

4. The following figures represent the time (in minutes) lost per day through mechanical failure of machinery collected over a period of 60 consecutive working days.

37	15	52	29	41	85	4	25	33	25	25	38
51	30	42	48	36	44	41	37	46	34	3	46
20	36	22	22	54	35	39	28	60	37	28	36
34	26	82	12	32	45	24	38	47	88	39	49
30	39	40	31	42	65	31	58	38	24	3	32

Construct a grouped frequency distribution and illustrate the date by drawing a histogram.

5. Use suitable diagrams to illustrate the following sets of data.

(a) Main method of transport to and from work

Type	Car	Train	Bus	Motorbike	Others
No. of employees	76	104	32	14	16

(b) Production figures in January 1982

Factory	A	B	C	D	E
No. of units produced	22 000	16 000	19 000	7000	12 000

(c) No. of employees in company A

Year	Age groups (years)			
	Under 25	25–35	35–45	Over 45
1970	1500	2200	1100	1200
1975	1400	2400	1400	1000
1980	900	1900	1400	1300

1.6 ARITHMETIC MEAN

One of the most important ways of describing a distribution is by stating the 'average'. One measure of an average is the arithmetic mean (denoted by $\bar{x}$).

Given a list of numbers
$$\bar{x} = \frac{\Sigma x}{n}$$

[Σ denotes 'the sum of'.]

Alternatively, in a frequency distribution
$$\bar{x} = \frac{\Sigma fx}{\Sigma f}$$

EXAMPLE 1.6.1 Calculate the arithmetic mean age of a group of employees whose ages (in years) are 21, 35, 18, 52, 48, 42.

SOLUTION
$$\bar{x} = \frac{\Sigma x}{n} = \frac{21 + 35 + 18 + 52 + 48 + 42}{6}$$

$$= \frac{216}{6} = 36$$

$\therefore$ mean age = 36 years

EXAMPLE 1.6.2 The number of industrial accidents over the past 60 working weeks are shown in the distribution below.

No. of accidents per week, x	0	1	2	3	4
No. of weeks, f	24	27	5	2	2

Calculate the arithmetic mean of the number of weekly accidents.

SOLUTION

x	0	1	2	3	4	
f	24	27	5	2	2	$\Sigma f = 60$
fx	0	27	10	6	8	$\Sigma fx = 51$

$$\bar{x} = \frac{\Sigma fx}{\Sigma f} = \frac{51}{60} = 0.85 \text{ accidents}$$

EXAMPLE 1.6.3 The number of units produced per day over the past 50 working days is shown in the table below.

Units produced	120-124	125-129	130-134	135-139	140-144	145-149
No. of days	1	0	8	26	13	2

Calculate the arithmetic mean number of units produced per day.

SOLUTION The value of x used is the midpoint of each class.

x	122	127	132	137	142	147	
f	1	0	8	26	13	2	$\Sigma f = 50$
fx	122	0	1056	3562	1846	294	$\Sigma fx = 6880$

$$\bar{x} = \frac{\Sigma fx}{\Sigma f} = \frac{6880}{50} = \underline{137.6 \text{ units per day}}$$

EXAMPLE 1.6.4 The age distribution of employees in a certain manufacturing company is shown below.

Age of employees (years)	16–	20–	25–	30–	40–	50–
No. of employees	23	53	81	109	73	61

Evaluate the arithmetic mean age of employees.

SOLUTION

x	18	22.5	27.5	35	45	57.5	
f	23	53	81	109	73	61	$\Sigma f = 400$
fx	414	1192.5	2227.5	3815	3285	3507.5	$\Sigma fx = 14\,441.5$

$$\bar{x} = \frac{\Sigma fx}{\Sigma f} = \frac{14\,441.5}{400} = \underline{36.1 \text{ years}}$$

1.7 THE MODE

The mode is another measure of the 'average', and is defined as being the most frequent value in the distribution.

EXAMPLE 1.7.1. The wheat yields in a particular region over the past 12 years are (in millions of tons): 1.3, 1.2, 1.0, 1.3, 1.4, 1.6, 1.7, 1.5, 1.3, 1.2, 1.1, 1.4.

Estimate the mode.

SOLUTION 1.3 occurs three times and is the most frequent value in this list. Therefore mode = $\underline{1.3 \text{ (million tons)}}$.

EXAMPLE 1.7.2 Estimate the mode for the distribution given in Example 1.6.2.

SOLUTION We have:

x	0	1	2	3	4
f	24	27	5	2	2

The highest frequency is 27, therefore mode = $\underline{1.}$

EXAMPLE 1.7.3   Estimate the mode for the distribution given in Example 1.6.3.

SOLUTION   We have:

x	120-124	125-129	130-134	135-139	140-144	145-149
f	1	0	8	26	13	2

The class 135-139 is the most frequent and is called the *modal class*. The mode is estimated from a histogram as shown below.

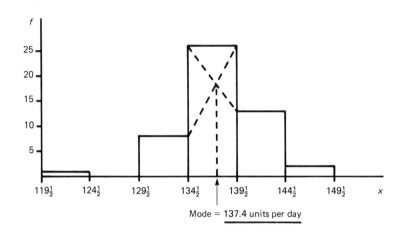

Mode = 137.4 units per day

EXAMPLE 1.7.4   Estimate the mode for the distribution given in Example 1.6.4.

SOLUTION   Now we have:

x	16-	20-	25-	30-	40-	50-
f	23	53	81	109	73	61
Class width	4	5	5	10	10	15
Frequency density	5.75	10.6	16.2	10.9	7.3	4.07

[*Note*: The mode occurs in the class represented by the highest block.]

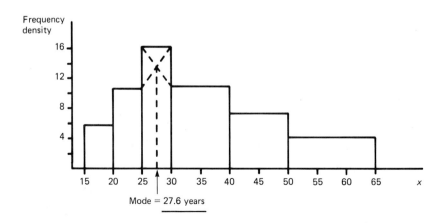

Mode = 27.6 years

1.8 MEDIAN

The median is a third method of evaluating the 'average' and is defined as being equal to the central value in a list of data given in numerical order. In general, the median is the $\frac{1}{2}(n+1)$th number in a list of n values.

EXAMPLE 1.8.1 The number of hours of overtime worked by a sample of nine employees are 2, 4, 5, 7, 3, 0, 2, 4, 8.

Find the median number of overtime hours.

SOLUTION Rearranging the data, we have: 0, 2, 2, 3, 4, 4, 5, 7, 8. The number of values is $n = 9$.

$$\text{Median} = \left(\frac{n+1}{2}\right)\text{th} = \left(\frac{9+1}{2}\right) = 5\text{th number}$$

$$= \underline{4} \text{ in this list}$$

EXAMPLE 1.8.2 Recalculate the median in Example 1.8.1 if a tenth employee works only 1 hour overtime.

SOLUTION We have: 0, 1, 2, 2, 3, 4, 4, 5, 7, 8, and $n = 10$.

$$\text{Median} = \left(\frac{n+1}{2}\right)\text{th} = \left(\frac{11}{2}\right) = 5\frac{1}{2}\text{th number}$$

i.e. midway between the 5th and 6th numbers. Therefore median = $\underline{3.5 \text{ hours}}$.

EXAMPLE 1.8.3 The total price of units ordered from a warehouse of a certain commodity is shown in the distribution below.

Cost of units ordered per day (£)	0–	50–	100–	150–	200–	250–
No. of days	2	7	8	16	9	8

Evaluate the median of this distribution.

SOLUTION We have $n = \Sigma f = 50$. Therefore

$$\text{median} = \left(\frac{50+1}{2}\right) = 25\frac{1}{2}\text{th number}$$

We estimate this value by drawing an ogive.

Cost (£)	Cumulative frequency
0	0
50	2
100	9
150	17
200	33
250	42
300	50

16   NOTES AND PROBLEMS IN STATISTICS

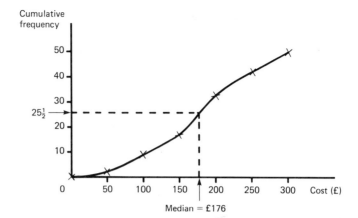

EXAMPLE 1.8.4   The time taken for the weekly maintenance of a group of machines in a workshop over the past 30 weeks is shown in the following table.

Maintenance time (hours)	Under 2	2–	4–	6–	10–	15–
No. of weeks	5	3	12	7	2	1

Estimate the median.

SOLUTION $n = \Sigma f = 30$, and

$$\text{median} = \left(\frac{30+1}{2}\right) = 15\tfrac{1}{2}\text{th number}$$

Time (h)	Cumulative frequency
0	0
2	5
4	8
6	20
10	27
15	29
20	30

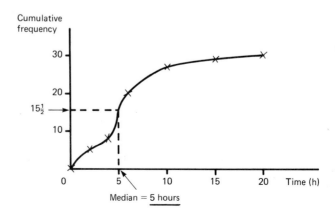

EXERCISES ON SECTIONS 1.6 TO 1.8

1. Estimate (a) the arithmetic mean, (b) the median and (c) the mode for both of the following frequency distributions.

(i)
x	0–	2–	4–	6–	8–
f	2	4	11	5	3

(ii)
x	10–	15–	20–	30–	50–
f	4	10	12	3	1

2. The following table gives the number of accidents occurring in each age group during 1981 in a large industrial complex.

Age group (years)	16–	20–	25–	30–	35–	40–	50–	60–
No. of accidents	6	13	14	12	9	11	9	4

Estimate the median and mode for this distribution.

3. The table below gives the number of units owned by a sample of 300 participants in a unit trust company.

No. of units	50–99	100–149	150–199	200–299
No. of participants	15	22	27	108

No. of units	300–399	400–499	500–999
No. of participants	60	48	20

Estimate the arithmetic mean of this distribution.

4. 180 batches of micro-chips were examined and the frequency distribution below gives the number of faulty chips in each batch.

No. of faulty chips	0–	5–	10–	15–	20–	25–	30–	35–	40–
No. of batches	25	38	44	29	18	12	8	4	2

Find the median number of faulty micro-chips per batch. Estimate the percentage of batches having more than 12 faulty micro-chips.

1.9 RANGE

Having quoted the 'average', a frequency distribution may also be described by a measure of spread (or dispersion). One such measure is the range. It is obtained by finding the difference between the largest and smallest values in the distribution.

EXAMPLE 1.9.1 Estimate the range for the distribution given in Example 1.8.1.

SOLUTION We have the values 2, 4, 5, 7, 3, 0, 2, 4, 8. Range = $8 - 0 = \underline{8 \text{ hours.}}$

EXAMPLE 1.9.2 Estimate the range for the distribution given in Example 1.8.4.

SOLUTION We have

x	0–	2–	4–	6–	10–	15–
f	5	3	12	7	2	1

In this distribution, minimum value = 0, maximum value = 20. Therefore, range = 20 − 0 = **20 hours**.

1.10 INTER-QUARTILE RANGE (IQR)

The IQR is a refinement of the range. It is the range containing the central 50% of the distribution, and is obtained by

$$IQR = Q_3 - Q_1$$

where

$$Q_1 = \text{lower quartile} = \left(\frac{n+1}{4}\right)\text{th number}$$

and

$$Q_3 = \text{upper quartile} = \frac{3(n+1)}{4}\text{th number}$$

Q_1 and Q_3 can be estimated from an ogive.

[*Note*: The *quartile deviation* = semi-inter-quartile range = $\frac{1}{2}(Q_3 - Q_1)$.]

EXAMPLE 1.10.1 The number of 'bad' cheques presented to a firm in the past 35 working days are shown in the table below. Estimate the inter-quartile range.

No. of bad cheques	0	1	2	3	4	5
No. of days	5	9	8	6	4	3

SOLUTION

$n = \Sigma f = 35$

$Q_1 = \left(\frac{35+1}{4}\right)\text{th} = $ 9th number = 1 from the distribution

$Q_3 = \frac{3}{4}(35+1)\text{th} = $ 27th number = 3 from the distribution

$\therefore \quad IQR = Q_3 - Q_1 = 3 - 1 = $ **2 cheques**

EXAMPLE 1.10.2 The numbers of telephone calls per day made from a particular department are shown in the following table. Estimate the inter-quartile range.

No. of calls	10–	20–	30–	40–	50–	60–	70–
No. of days	19	27	34	21	3	1	2

SOLUTION

$n = \Sigma f = 107$

$Q_1 = \left(\dfrac{107+1}{4}\right)$th = 27th number

$Q_3 = \frac{3}{4}(107+1)$th = 81st number

Q_1 and Q_3 are estimated from the ogive shown below.

No. of calls	Cumulative frequency
10	0
20	19
30	46
40	80
50	101
60	104
70	105
80	107

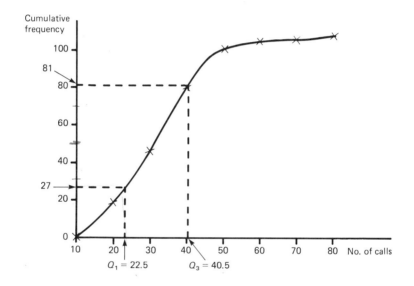

IQR = $Q_3 - Q_1$ = 40.5 − 22.5 = <u>18 calls</u>

EXAMPLE 1.10.3 The lengths of service of the last 109 employees to leave a firm are given below. Estimate the inter-quartile range.

Length of service (years)	Under 1	1–	5–	10–	15–	20–
No. of employees	14	36	28	19	5	7

SOLUTION

$n = \Sigma f = 109$

$Q_1 = \left(\dfrac{109+1}{4}\right) = 27\frac{1}{2}$th number

$Q_3 = \frac{3}{4}(109+1) = 82\frac{1}{2}$th number

Service (years)	Cumulative frequency
0	0
1	14
5	50
10	78
15	97
20	102
30	109

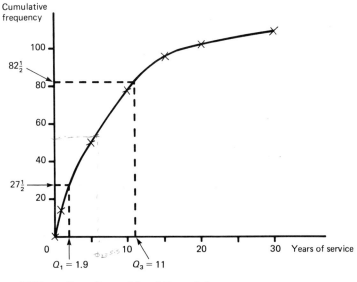

$$\text{IQR} = Q_3 - Q_1 = 11 - 1.9 = \underline{9.1 \text{ years}}$$

1.11 MEAN DEVIATION

The mean deviation is a further measure of 'spread' and is the average distance between every value in the distribution and the arithmetic mean, $\bar{x}$.

Given a list of values, $$\text{mean deviation} = \frac{\Sigma|x - \bar{x}|}{n}$$

Alternatively in a frequency distribution, $$\text{mean deviation} = \frac{\Sigma f|x - \bar{x}|}{\Sigma f}$$

EXAMPLE 1.11.1 The ages (in years) of a random sample of 10 customers are 19, 41, 23, 36, 17, 22, 51, 26, 34, 31.

Calculate the mean deviation.

SOLUTION We have x: 19, 41, 23, 36, 17, 22, 51, 26, 34, 31. Now

$$\bar{x} = \frac{\Sigma x}{n} = \frac{300}{10} = 30$$

We have $|x-\bar{x}|$: 11, 11, 7, 6, 13, 8, 21, 4, 4, 1.

$$\text{Mean deviation} = \frac{\Sigma|x-\bar{x}|}{n} = \frac{86}{10} = 8.6 \text{ years}$$

EXAMPLE 1.11.2 Estimate the mean deviation of the number of errors detected in a random sample of 50 accounts.

No. of errors	0	1	2	3	4
No. of accounts	18	23	6	2	1

SOLUTION

x	f	fx	$\|x-\bar{x}\|$	$f\|x-\bar{x}\|$
0	18	0	0.9	16.2
1	23	23	0.1	2.3
2	6	12	1.1	6.6
3	2	6	2.1	4.2
4	1	4	3.1	3.1
	$\Sigma f = 50$	$\Sigma fx = 45$		$\Sigma f\|x-\bar{x}\| = 32.4$

$$\bar{x} = \frac{\Sigma fx}{\Sigma f} = \frac{45}{50} = 0.9$$

$$\text{Mean deviation} = \frac{\Sigma f|x-\bar{x}|}{\Sigma f} = \frac{32.4}{50} = 0.648 \text{ errors}$$

EXAMPLE 1.11.3 Find the mean deviation of the distribution of basic hourly wages given in the following table.

Hourly wage (£)	1.20–	1.40–	1.60–	1.80–	2.00–	2.20–	2.40–
Percentage of earners	6	8	16	36	23	7	4

SOLUTION The value of x is assumed to be the midpoint of each class.

x	f	fx	$\|x-\bar{x}\|$	$f\|x-\bar{x}\|$
1.3	6	7.8	0.598	3.588
1.5	8	12	0.398	3.184
1.7	16	27.2	0.198	3.168
1.9	36	68.4	0.002	0.072
2.1	23	48.3	0.202	4.646
2.3	7	16.1	0.402	2.814
2.5	4	10	0.602	2.408
	$\Sigma f = 100$	$\Sigma fx = 189.8$		$\Sigma f\|x-\bar{x}\| = 19.88$

$$\bar{x} = \frac{\Sigma fx}{\Sigma f} = \frac{189.8}{100} = 1.898$$

$$\text{Mean deviation} = \frac{\Sigma f|x-\bar{x}|}{\Sigma f} = \frac{19.88}{100} = 0.1988 \quad \text{(approx. £0.20)}$$

1.12 STANDARD DEVIATION

The standard deviation (denoted by s) is a refinement of the mean deviation and is considered to be the most important measure of spread.

Given a list of values
$$s = \sqrt{\frac{\Sigma(x-\bar{x})^2}{n}}$$

Alternatively, in a frequency distribution,
$$s = \sqrt{\frac{\Sigma f(x-\bar{x})^2}{\Sigma f}}$$

Such formulae can be simplified. In particular in a frequency distribution the revised formula for s can be written as:
$$s = \sqrt{\frac{\Sigma fx^2}{\Sigma f} - \left(\frac{\Sigma fx}{\Sigma f}\right)^2}$$

[Note: The value of s^2 is called the *variance*.]

$$\text{Variance} = \frac{\Sigma fx^2}{\Sigma f} - \left(\frac{\Sigma fx}{\Sigma f}\right)^2$$

EXAMPLE 1.12.1 Find the standard deviation of the distribution given in Example 1.11.1.

SOLUTION We have x: 19, 41, 23, 36, 17, 22, 51, 26, 34, 31.

$$\bar{x} = \frac{\Sigma x}{n} = \frac{300}{10} = 30$$

We have:

$(x-\bar{x})$	−11	11	−7	6	−13	−8	21	−4	4	1
$(x-\bar{x})^2$	121	221	49	36	169	64	441	16	16	1

$$\Sigma(x-\bar{x})^2 = 1034$$

$$\text{Standard deviation, } s = \sqrt{\frac{\Sigma(x-\bar{x})^2}{n}} = \sqrt{\frac{1034}{10}} = \sqrt{103.4}$$

$$= \underline{10.17 \text{ years}}$$

EXAMPLE 1.12.2 Find the standard deviation of the distribution given in Example 1.11.2.

SOLUTION The distribution is shown below.

x	f	fx	fx^2
0	18	0	0
1	23	23	23
2	6	12	24
3	2	6	18
4	1	4	16
	$\Sigma f = 50$	$\Sigma fx = 45$	$\Sigma fx^2 = 81$

$$s = \sqrt{\frac{\Sigma fx^2}{\Sigma f} - \left(\frac{\Sigma fx}{\Sigma f}\right)^2} = \sqrt{\frac{81}{50} - \left(\frac{45}{50}\right)^2}$$

$$= \sqrt{1.62 - 0.81}$$

$$= \sqrt{0.81}$$

Standard deviation = 0.9 errors

EXAMPLE 1.12.3 Find the standard deviation of the distribution given in Example 1.11.3.

SOLUTION The calculations are shown in the table below.

x	f	fx	fx^2
1.3	6	7.8	10.14
1.5	8	12.0	18.00
1.7	16	27.2	46.24
1.9	36	68.4	129.96
2.1	23	48.3	101.43
2.3	7	16.1	37.03
2.5	4	10.0	25.00
	$\Sigma f = 100$	$\Sigma fx = 189.8$	$\Sigma fx^2 = 367.80$

$$s = \sqrt{\frac{\Sigma fx^2}{\Sigma f} - \left(\frac{\Sigma fx}{\Sigma f}\right)^2} = \sqrt{\frac{367.8}{100} - \left(\frac{189.8}{100}\right)^2}$$

$$= \sqrt{3.678 - 3.6024}$$

$$= \sqrt{0.0756}$$

Standard deviation = 0.275 (approx. £0.27)

1.13 CODING

To simplify calculations we can use the transformation

$$X = \frac{x - a}{b}$$

We then find $\bar{X}$ and s_X and use the formulae

$$\bar{x} = b\bar{X} + a$$

and

$$s_x = bs_X$$

EXAMPLE 1.13.1 Find the mean and standard deviation of the distribution of annual salaries in the sample given below.

Salary (£)	4000–	5500–	7000–	8500–	10 000–
No. of employees	8	14	27	16	15

SOLUTION

x	f	$x-4750$	$X = \dfrac{x-4750}{1500}$	fX	fX^2
4750	8	0	0	0	0
6250	14	1500	1	14	14
7750	27	3000	2	54	108
9250	16	4500	3	48	144
10 750	15	6000	4	60	240
	$\Sigma f = 80$			$\Sigma fX = 176$	$\Sigma fX^2 = 506$

$$\therefore \quad \overline{X} = \frac{\Sigma fX}{\Sigma f} = \frac{176}{80} = 2.2$$

and

$$s_X = \sqrt{\frac{\Sigma fX^2}{\Sigma f} - \left(\frac{\Sigma fX}{\Sigma f}\right)^2}$$

$$= \sqrt{\frac{506}{80} - \left(\frac{176}{80}\right)^2}$$

$$= \sqrt{6.325 - 4.84}$$

$$= \sqrt{1.485}$$

$$= \underline{1.2186}$$

$$\therefore \quad \bar{x} = 1500\overline{X} + 4750 = 1500(2.2) + 4750$$

$$= 8050$$

and

$$s_x = 1500 s_X = 1500(1.2186)$$

$$= 1827.9$$

$$\therefore \quad \text{Mean} = \underline{£8050}, \quad \text{Standard deviation} = \underline{£1827.90}$$

EXERCISES ON SECTIONS 1.9 TO 1.13

1. Find (a) the inter-quartile range, (b) the mean deviation and (c) the standard deviation for both of the following frequency distributions.

(i)

x	0–	6–	12–	18–	24–
f	1	2	4	1	1

(ii)

x	6–	10–	14–	20–	30–
f	5	11	24	7	3

2. The price of a specific electrical item in a random sample of 40 retail establishments over the UK is shown below.

Price (£)	90–	100–	110–	120–	130–	140–
No. of stores	3	2	26	3	2	4

Find the arithmetic mean and standard deviation of this distribution.

3. The table below gives the size of the last 120 orders received by a mail-order company.

Size of order (£)	0–	5–	10–	15–	20–	30–	40–	60–
No. of orders	11	21	26	22	18	12	5	5

Estimate the median and inter-quartile range for this distribution.

4. A production department in a manufacturing company had the following wastage among semi-skilled personnel during 1980/81.

Employment period (months)	0–	3–	6–	12–	24–	36–
No. of leavers	25	32	16	12	7	8

Find the mean and standard deviation of this distribution.

5. Find the arithmetic mean and standard deviation of the daily production figures given in the following table.

Production (no. of units)	1200–	1235–	1270–	1305–	1340–
No. of days	12	21	14	6	7

PAST QUESTIONS 1

1. Your company manufactures an integrated circuit chip. The operating life times of a sample of 100 chips from the current output were recorded as follows:

2000 to less than 3000 hours	3 chips
3000 to less than 4000 hours	13 chips
4000 to less than 5000 hours	19 chips
5000 to less than 6000 hours	27 chips
6000 to less than 7000 hours	21 chips
7000 to less than 8000 hours	12 chips
8000 to less than 9000 hours	5 chips

(a) Calculate the mean and standard deviation of the operating life times of these 100 chips.

(b) The company offers a free replacement guarantee on all chips failing within the first 3200 hours of operating life. Using the above frequency distribution, estimate the percentage of chips that would need to be replaced. [IIM]

2. Using the frequency distribution given below:
   (a) Draw a histogram and estimate the mode.
   (b) Draw a cumulative frequency curve and estimate the median.

Time slot (mins)	Frequency
0 under 10	10
10 under 30	20
30 under 60	60
60 under 100	60
100 under 150	50

   [IMS]

3. In a test to determine the working life of a type of electric light bulb, 100 bulbs were selected at random from production and simultaneously connected to a power source. The following data show the number out of the original 100 still working at the end of successive periods of 100 hours, all bulbs having failed within 1000 hours.

Elapsed time (hours)	No. working
0	100
100	99
200	98
300	90
400	82
500	70
600	45
700	26
800	12
900	3
1000	0

   Required:
   (a) By graphing these data, estimate:
      (i) the median working life
      (ii) the interquartile range of working life
      for this type of bulb (Do not attempt to use mathematical formulae.)
   (b) Show how to deduce the frequency distribution of working life of the 100 bulbs from the above data and use your distribution to estimate the arithmetic mean working life of a bulb.
   (c) Suppose that a bulb has already been working for 700 hours. Calculate how many more hours of operation could be expected from it. [ACA]

4. (a) Using the figures given on the page opposite calculate:
      (i) the range
      (ii) the arithmetic mean
      (iii) the median
      (iv) the lower quartile
      (v) the upper quartile
      (vi) the quartile deviation
      (vii) the mean deviation
      (viii) the standard deviation.

2	15	26	39	47	58
5	17	30	40	51	60
7	18	32	43	53	64
8	22	36	45	55	66
11					

(b) Explain the term 'measure of dispersion' and state briefly the advantages of using the following measures of dispersion:
  (i) range
  (ii) quartile deviation
  (iii) mean deviation
  (iv) standard deviation. [ICMA]

5. (a) Illustrate the following data by means of a histogram, a frequency polygon, and a cumulative frequency polygon (ogive).

Weekly wage (£)	No. of employees
31 and < 36	6
36 and < 41	8
41 and < 46	12
46 and < 51	18
51 and < 56	25
56 and < 61	30
61 and < 66	24
66 and < 71	14
71 and < 76	6
76 and < 81	3

(b) A large bank, in its accounts for three years, showed its profit distribution as follows:

	1978 (£m)	1977 (£m)	1976 (£m)
Profit before taxation	373	295	198
Taxation	135	140	82
Minority interests	12	12	11
Dividends	30	23	20
Retained profit	196	120	85

You are required to present the information in a visual display form. [ICMA]

6. Weekly exports of mechanical excavators

132	93	85	77	108	63	115	93	100	110	70
71	83	126	85	79	99	81	83	57	99	77
92	95	53	76	92	65	112	88	90	96	102
105	89	101	65	84	104	80	98	91	87	100
58	75	72	100	88	75	103	65	104	95	90
116	64	90	102	79	99	84	96	99	117	83
91	89	98	87	75	81	98	102	85	77	76
123	104	69	94	107	111	136	78	64	71	83

For the above set of figures:

(a) Construct a suitably grouped frequency table.

(b) Calculate the arithmetic mean from the original 88 readings and also from the grouped frequency table; comment on any difference.

(c) Calculate the standard deviation from the grouped frequency table. [IM]

7. A large company monitors its weekly sales levels in two different overseas markets and concludes there is no seasonal pattern to be found. However, the sales distribution for the two markets is shown in the following table.

Sales level (£ thousands)	No. of weeks at given sales level	
	Market A	Market B
6 but less than 8	1	4
8 but less than 10	3	6
10 but less than 12	3	7
12 but less than 14	4	9
14 but less than 16	12	11
16 but less than 18	26	8
18 but less than 20	3	7
	52	52

(a) Calculate the arithmetic mean and standard deviation sales level for the two market areas.

(b) In the light of these results, discuss the nature and significance of these two markets to the company. [IM]

8. The fares collected by a bus company from 200 000 passengers are shown in the following frequency distribution.

Fare (pence)	No. collected (÷1000)
6	9
10	22
12	37
15	52
19	31
22	22
27	14
32	8
38	5
	200

The company issues tickets of denomination 1 pence, 2 pence, 5 pence and 13 pence. The minimum number of tickets is given to a passenger when the fare is collected so that, for example, a 27 pence fare is given two 13 pence and one 1 pence tickets.

Required:

(a) Construct two frequency distributions, one being the distribution of the number of tickets issued per fare and the other the distribution of the ticket denominations issued.

(b) Use the distributions in part (a) to obtain:
   (i) the mean number of tickets issued per fare
   (ii) the median number of tickets issued per fare
   (iii) the modal number of tickets issued per fare
   (iv) the percentage of tickets issued which are 5 pence denomination.
   [ACA]

9. The following data show scrap losses in kilos from two steel processes producing the same finished items:

Process A					Process B				
48	22	40	6	30	6	14	54	18	4
26	73	21	8	34	15	21	10	6	15
3	6	26	15	31	18	50	3	3	18
20	13	27	18	6	13	7	48	30	7
11	10	14	21	13	8	14	19	63	22
14	15	9	62	44	3	17	20	84	19
49	58	8	7	29	3	4	4	7	23
32	7	1	4	8	10	34	7	13	8
5	14	10	29	11	30	19	7	18	12
30	23	9	8	6	9	55	20	28	11

(a) Construct separate grouped frequency distributions for the two data sets using the same class intervals in each case.

(b) Plot the percentage ogives for the two distributions on the same graph.

(c) By determining suitable summary statistics from the ogives, contrast the two processes with respect to their effect on scrap loss.   [IIM]

10. A machine produces the following number of rejects in each successive period of five minutes:

16	21	26	24	11	24	17	25	26	13
27	24	26	3	27	23	24	15	22	22
12	22	29	21	18	22	28	25	7	17
22	28	19	23	23	22	3	19	13	31
23	28	24	9	20	33	30	23	20	8

(a) Construct a frequency distribution from these data, using seven class intervals of equal width.

(b) Using the frequency distribution, calculate an appropriate measure of:
   (i) average
   (ii) dispersion.

(c) Briefly explain the meaning of your calculated measures.   [ICMA]

11. The table shows the academic attainment of girls leaving school in England and Wales, expressed as percentages for the two years stated.

(a) Construct accurately a diagram to illustrate best the changes that have occurred between 1966 and 1976.

(b) Construct accurately, showing any working, a pie diagram to illustrate the statistics for 1976.

Attainment	1965–66	1975–76
3 or more 'A' level passes	4.8	7.0
2 'A' level passes	3.8	4.5
1 'A' level pass	2.8	3.3
5 or more higher grade 'O' level or CSE grade 1 passes	9.6	7.4
1–4 such passes	16.4	28.3
1 or more other grade GCE or CSE passes	37.1	32.5
No GCE or CSE qualifications	25.5	17.0

(*Source*: Department of Education and Science)   [IM]

12. The income distribution of the middle management in a large organisation is tabulated below:

Incomes	
Range (£)	No. of managers
5900 and less than 6300	10
6300 and less than 6700	8
6700 and less than 7100	18
7100 and less than 7500	22
7500 and less than 7900	23
7900 and less than 8300	25
8300 and less than 8700	35
8700 and less than 9100	30
9100 and less than 9500	14
9500 and less than 9900	6
9900 and less than 10 300	5
10 300 and less than 10 700	4

Draw a cumulative frequency curve of the above distribution, calculate the first and third quartile salaries and the median salary and show these on the graph. Write a brief note for your managing director on the implications of the analysis for wages policy.   [IPM]

13. The table expresses the number of employees in a selection of firms operating in a particular sector of the market in the USA and Japan.

Construct a cumulative frequency curve for each country and hence, by interpolation from the graphs, estimate the values of the median and quartiles and calculate the quartile deviation in each case. Discuss the implications of your results.

No. of employees	No. of firms	
	USA	Japan
Under 11	5	54
11 to 20	11	49
21 to 35	18	31
36 to 60	29	16
61 to 100	42	8
101 to 150	67	2

[IM]

FREQUENCY DISTRIBUTIONS   31

14. The price of the Ordinary 25 p Shares of Manco plc quoted on the Stock Exchange at the close of business on successive Fridays is tabulated below.

126	120	122	105	129	119	131	138
125	127	113	112	130	122	134	136
128	126	117	114	120	123	127	140
124	127	114	111	116	131	128	137
127	122	106	121	116	135	142	130

Required:

(a) Group the above data into eight classes.

(b) By constructing the ogive calculate the median value, quartile values and the semi-interquartile range.

(c) Calculate the mean and standard deviation of your frequency distribution.

(d) Compare and contrast the values that you have obtained for:
   (i) the median and mean
   (ii) the semi-interquartile range and the standard deviation.   [ACA]

15. An organisation is investigating the various absenteeism hours per employee over a particular period. A preliminary sample of 40 cards gave the following results.

34.5	37.0	31.0	33.5	48.0	40.0	33.0	43.5
42.5	40.0	33.0	39.5	39.5	31.5	40.5	34.0
38.0	34.5	28.0	41.5	35.0	34.0	37.0	37.0
31.5	37.0	29.0	38.0	31.0	23.5	29.0	41.0
34.5	30.5	37.5	32.5	38.0	39.5	44.5	36.5

(a) Construct a grouped frequency distribution for the data.

(b) What important measures could you determine from the grouped frequency distribution?

(c) Describe, very briefly, what these measures mean.   [ITD]

16. The following figures detail the premium income in £000's received by 30 branches of an insurance company.

100	70	730	400	50
270	420	400	630	880
420	510	310	130	740
310	90	520	540	220
1350	970	310	810	190
920	620	1210	720	540

The figures have been rounded to the nearest £10 000 in each case.

(a) Calculate the mean of this distribution using a grouped frequency table, making clear any assumptions that are involved.

(b) Calculate the mean deviation of the income figures, from the table.

(c) How would you compare the use of this figure with other measures of dispersion?   [CII]

17. (a) Define and describe three different averages indicating when they could be used.

(b) Items are manufactured to the same nominal length on two different machines A and B. The results of tests on items from each machine are given below.

Class midpoints (mm)	20	22	24	26	28	30
Frequency machine A	4	6	20	50	14	6
Frequency machine B	12	28	30	18	12	–

Calculate an average for both sets of data and state whether you think the machines are producing items of the same average length.

Draw a cumulative frequency diagram for both sets of data on the same graph.

If items below 22 mm are too short and are scrapped and replaced at a cost of 50p and items over 28 mm are too long and are reworked at a cost of 30p each, what is the total extra cost to produce 100 perfect items on each machine? Which machine would you recommend to use for the production of the items? [RICS]

2 PROBABILITY

2.1 BASIC PROBABILITY

The probability of an event X is

$$p(X) = \frac{\text{number of ways X can occur}}{\text{total number of possible outcomes}}$$

The probability of X not occurring is

$$p(\overline{X}) = 1 - p(X)$$

EXAMPLE 2.1.1 Find the probability of obtaining a club when taking one card from a pack.

SOLUTION $$p(\text{club}) = \frac{13}{52} = \underline{\frac{1}{4}} \quad (or\ 0.25)$$

EXAMPLE 2.1.2 A group consists of 30 males and 20 females. One member of this group is selected at random. Find the probability that a male is selected.

SOLUTION $$p(\text{male}) = \frac{30}{50} = \underline{\frac{3}{5}} \quad (or\ 0.6)$$

EXAMPLE 2.1.3 If 65% of questionnaires distributed in a survey are returned completed, what is the chance that a questionnaire (chosen at random) is not returned?

SOLUTION $p(X)$ $p(\text{return}) = \dfrac{65}{100} = 0.65\ (65\%)$

$\therefore\ p(\overline{X})$ $p(\text{not return}) = 1 - p(\text{return})$

$\phantom{\therefore\ p(\overline{X})\ p(\text{not return})} = 1 - 0.65$

$\phantom{\therefore\ p(\overline{X})\ p(\text{not return})} = \underline{0.35}\ (35\%)$

2.2 COMBINATION OF EVENTS

$p(X \text{ and } Y) = p(X) \cdot p(Y)$ if X and Y are independent events

$p(X \text{ or } Y) = p(X) + p(Y)$ if X and Y are mutually exclusive events

EXAMPLE 2.2.1 If X and Y are mutually exclusive events with $p(X) = 0.1$ and $p(Y) = 0.3$ find $p(X \text{ or } Y)$.

SOLUTION

$$p(X \text{ or } Y) = p(X) + p(Y)$$
$$= 0.1 + 0.3$$
$$= \underline{0.4}$$

EXAMPLE 2.2.2 If X and Y are independent events with $p(X) = 0.3$ and $p(Y) = 0.6$ find:
(a) $p(X \text{ and } Y)$
(b) $p(\overline{X} \text{ and } Y)$.

SOLUTION (a)
$$p(X \text{ and } Y) = p(X) \cdot p(Y)$$
$$= (0.3)(0.6)$$
$$= \underline{0.18}$$

(b) $p(\overline{X}) = 1 - p(X) = 1 - 0.3 = 0.7$

∴ $p(\overline{X} \text{ and } Y) = p(\overline{X}) \cdot p(Y)$
$$= (0.7)(0.6)$$
$$= \underline{0.42}$$

EXAMPLE 2.2.3 20% of employees in a company earn over £150 per week and 50% earn between £100 and £150 per week. Find the probability that an employee selected at random earns:
(a) less than £100 per week
(b) under £100 or over £150 per week.

SOLUTION
$$p(\text{over } £150) = \frac{20}{100} = 0.2$$

$$p(£100 \text{ to } £150) = \frac{50}{100} = 0.5$$

(a) $p(\text{under } £100) = 1 - p(\text{over } £100)$
$$= 1 - [p(£100 \text{ to } £150) + p(\text{over } £150)]$$
$$= 1 - [0.5 + 0.2]$$
$$= 1 - 0.7$$
$$= \underline{0.3} \quad (30\%)$$

(b) $p(\text{under } £100 \text{ or over } £150) = p(\text{under } £100) + p(\text{over } £150)$
$$= 0.3 + 0.2$$
$$= \underline{0.5} \quad (50\%)$$

EXAMPLE 2.2.4 Students take two independent tests. 30% of students pass test A and 60% pass test B. Find the probability that a student selected at random passes:
(a) both tests
(b) only test A
(c) only one test.

SOLUTION $p(\text{A pass}) = \dfrac{30}{100} = 0.3, \quad p(\text{B pass}) = \dfrac{60}{100} = 0.6$

(a) $p(\text{both passes}) = p(\text{A pass and B pass})$
$= p(\text{A pass}) \cdot p(\text{B pass})$
$= (0.3)(0.6)$
$= \underline{0.18}$ (i.e. 18% of students pass both tests)

(b) $p(\text{only A pass}) = p(\text{A pass and B fail})$
$= p(\text{A pass}) \cdot p(\text{B fail})$
$= (0.3)(1 - 0.6)$
$= (0.3)(0.4)$
$= \underline{0.12}$ (12%)

(c) $p(\text{only one pass}) = p(\text{A pass and B fail or A fail and B pass})$
$= p(\text{A pass}) \cdot p(\text{B fail}) + p(\text{A fail}) \cdot p(\text{B pass})$
$= (0.3)(0.4) + (0.7)(0.6)$
$= 0.12 + 0.42$
$= \underline{0.54}$ (54%)

EXERCISES ON SECTIONS 2.1 AND 2.2

1. Given that A and B are independent events with $p(A) = 0.1$ and $p(B) = 0.7$, find the following probabilities:
   (a) $p(\bar{A})$
   (b) $p(\bar{B})$
   (c) $p(A \text{ and } B)$
   (d) $p(\bar{A} \text{ and } \bar{B})$.

2. Given that X and Y are mutually exclusive events with $p(X) = 0.4$ and $p(Y) = 0.5$, find the following probabilities:
   (a) $p(\bar{X})$
   (b) $p(\bar{Y})$
   (c) $p(X \text{ or } Y)$.

3. It is known that approximately 10% of items produced on an assembly line are defective. If two items are taken at random, what is the probability that:
   (a) neither item is defective
   (b) only one item is defective
   (c) both items are defective?

4. A salesman claims that one in every three of his contacts with clients results in a sale. If, on a given day, he contacts two clients, what is the chance that he obtains:

(a) no sales,

(b) only one sale?

5. 20% of trainees with a company fail to complete their training period. In a random sample of three trainees what is the chance that:

(a) they all complete the training period

(b) only one trainee completes the training period?

6. In a particular manufacturing industry 30% of employees work over 10 hours per week overtime, whereas 45% do no overtime. In a sample of two employees what is the probability that:

(a) both employees do less than 10 hours per week overtime

(b) only one does no overtime?

2.3 COMBINATIONS

$$n \text{ factorial} = n! = n(n-1)(n-2) \ldots 3 \times 2 \times 1$$

The number of combinations of n items taken r at a time is given by:

$$^nC_r = \frac{n!}{r!(n-r)!}$$

$$\left[\text{alternative notation is } \binom{n}{r} \right]$$

EXAMPLE 2.3.1 Find the following factorials: (a) 3!, (b) 6!, (c) 1!, (d) 0!.

SOLUTION (a) $3! = 3 \times 2 \times 1 = \underline{6}$

(b) $6! = 6 \times 5 \times 4 \times 3 \times 2 \times 1 = \underline{720}$

(c) $1! = \underline{1}$

(d) $0! = \underline{1}$ [NB: a special case]

EXAMPLE 2.3.2 Evaluate the following expressions involving factorials:

(a) $\dfrac{6!}{4!}$ (b) $2!4!$ (c) $\dfrac{10!}{8!}$ (d) $\dfrac{20!}{18!2!}$.

SOLUTION (a) $\dfrac{6!}{4!} = \dfrac{6 \times 5 \times \cancel{4 \times 3 \times 2 \times 1}}{\cancel{4 \times 3 \times 2 \times 1}} = \underline{30}$

(b) $2!4! = (2 \times 1)(4 \times 3 \times 2 \times 1) = (2)(24) = \underline{48}$

(c) $\dfrac{10!}{8!} = \dfrac{10 \times 9 \times \cancel{8 \times 7 \times 6 \times 5 \times 4 \times 3 \times 2 \times 1}}{\cancel{8 \times 7 \times 6 \times 5 \times 4 \times 3 \times 2 \times 1}} = \underline{90}$

(d) $\dfrac{20!}{18!2!} = \dfrac{20 \times 19 \times \cancel{18 \times 17 \ldots 3 \times 2 \times 1}}{\cancel{(18 \times 17 \ldots 3 \times 2 \times 1)}(2 \times 1)} = \dfrac{380}{2} = \underline{190}$

EXAMPLE 2.3.3 Find the following number of combinations: (a) $^5C_3$, (b) $^3C_0$, (c) $^{20}C_{18}$.

SOLUTION (a) $^5C_3 = \dfrac{5!}{3!(5-3)!} = \dfrac{5!}{3!2!} = \dfrac{5 \times 4 \times \cancel{3 \times 2 \times 1}}{\cancel{(3 \times 2 \times 1)}(2 \times 1)} = \dfrac{20}{2} = \underline{10}$

(b) $^3C_0 = \dfrac{3!}{0!3!} = \dfrac{3 \times 2 \times 1}{1(3 \times 2 \times 1)} = \underline{1}$

(c) $^{20}C_{18} = \dfrac{20!}{18!2!} = \underline{190}$ (see Example 2.3.2)

EXAMPLE 2.3.4 Find the number of different ways of selecting a working party of 3 employees from a group containing 10 members.

SOLUTION Number of ways of selecting 3 from 10 is

$$^{10}C_3 = \dfrac{10!}{3!7!} = \dfrac{10 \times 9 \times 8 \times \cancel{7 \times 6 \ldots 2 \times 1}}{(3 \times 2 \times 1)\cancel{(7 \times 6 \ldots 2 \times 1)}} = \dfrac{720}{6} = \underline{120}$$

i.e. there are 120 different ways of selecting the working party.

2.4 BINOMIAL DISTRIBUTION

If p is the probability of a 'success' in an individual trial, then in n independent trials the probability of r successes is given by the Binomial formula:

$$p(r) = {^nC_r} \cdot p^r \cdot (1-p)^{n-r}$$

where $r = 0, 1, 2, \ldots, n$. The number of successes has a mean $= np$ and standard deviation $= \sqrt{np(1-p)}$.

EXAMPLE 2.4.1 If $n = 4$ and $p = 0.2$, use the Binomial formula to find: (a) $p(3)$, (b) $p(0)$.

SOLUTION (a) $p(3) = {^4C_3}(0.2)^3(1-0.2)^{4-3}$

$= \dfrac{4!}{3!1!}(0.2)^3(0.8)^1$ $(0.85)\ (0.15)$

$= 4(0.008)(0.8)$

$\therefore \quad p(3) = \underline{0.0256}$

(b) $p(0) = {^4C_0}(0.2)^0(0.8)^4$

$= \dfrac{4!}{0!4!}(1)(0.4096)$

$\therefore \quad p(0) = \underline{0.4096}$

EXAMPLE 2.4.2 By drawing a histogram illustrate the Binomial distribution where $n = 4$ and $p = 0.5$.

SOLUTION
$p(0) = {}^4C_0(0.5)^0(0.5)^4 = 0.0625$
$p(1) = {}^4C_1(0.5)^1(0.5)^3 = 0.25$
$p(2) = {}^4C_2(0.5)^2(0.5)^2 = 0.375$
$p(3) = {}^4C_3(0.5)^3(0.5)^1 = 0.25$
$p(4) = {}^4C_4(0.5)^4(0.5)^0 = 0.0625$

These probabilities are illustrated in the histogram shown below.

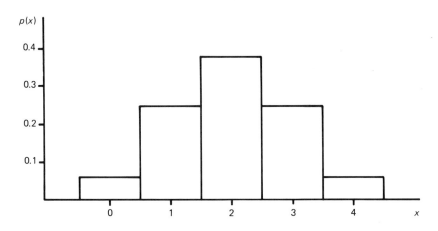

EXAMPLE 2.4.3 From past experience it is known that approximately 60% of applicants pass an initial assessment test. In a group of five applicants find the probability that:

(a) all applicants pass the test
(b) only three applicants pass
(c) more than three applicants pass.

SOLUTION We have $n = 5$, $p = 0.6$ (60%).

(a) $\quad p(5) = {}^5C_5(0.6)^5(0.4)^0$

$\quad\quad = \dfrac{5!}{5!0!}(0.07776)(1)$

$\quad\quad = \underline{0.07776}\quad$ (an almost 8% chance)

(b) $\quad p(3) = {}^5C_3(0.6)^3(0.4)^2$

$\quad\quad = \dfrac{5!}{3!2!}(0.216)(0.16)$

$\quad\quad = 10(0.216)(0.16)$

$\quad\quad = \underline{0.3456}\quad (35\%)$

(c) $p(\text{more than } 3) = p(4 \text{ or } 5) = p(4) + p(5)$

Now $p(4) = {}^5C_4(0.6)^4(0.4)^1$

$= 5(0.1296)(0.4)$

$p(4) = 0.2592$

and $p(5) = 0.07776$ (from part (a))

$\therefore \quad p(\text{more than } 3) = 0.2592 + 0.07776$

$= \underline{0.33696} \quad (34\%)$

EXAMPLE 2.4.4 One-quarter of all accounts are found to contain errors. In a batch of eight accounts find the probability that the number of accounts containing errors is:

(a) less than two

(b) more than two.

Find the mean and standard deviation of the number of accounts containing errors.

SOLUTION We have $n = 8$, $p = \frac{1}{4}(0.25)$.

(a) $p(\text{less than } 2) = p(0) + p(1)$

Now $p(0) = {}^8C_0(\frac{1}{4})^0(\frac{3}{4})^8 = 0.10011$

and $p(1) = {}^8C_1(\frac{1}{4})^1(\frac{3}{4})^7 = 0.26697$

$\therefore \quad p(\text{less than } 2) = 0.10011 + 0.26697$

$= \underline{0.36708}$

(b) $p(\text{more than } 2) = p(3 \text{ or } 4 \text{ or } 5 \text{ or} \ldots)$

$= 1 - [p(0) + p(1) + p(2)]$

Now $p(2) = {}^8C_2(\frac{1}{4})^2(\frac{3}{4})^6 = 0.31146$

$\therefore \quad p(\text{more than } 2) = 1 - [0.10011 + 0.26697 + 0.31146]$

$= 1 - 0.67854$

$= \underline{0.32146}$

Mean $= np = 8 \times \frac{1}{4} = \underline{2}$

Standard deviation $= \sqrt{np(1-p)} = \sqrt{8(\frac{1}{4})(\frac{3}{4})} = \sqrt{1.5} = \underline{1.2247}$

2.5 POISSON DISTRIBUTION

If events occur at random at an average rate of λ per unit time then the probability of r events is given by the Poisson formula $\boxed{p(r) = \dfrac{e^{-\lambda} \cdot \lambda^r}{r!}}$

where $r = 0, 1, 2, \ldots$

Mean $= \lambda$ Standard deviation $= \sqrt{\lambda}$

The Poisson distribution can be used to approximate the Binomial distribution when n is large $(n > 30)$ and p is small $(p < 0.1)$.

EXAMPLE 2.5.1 When $\lambda = 2$, find the Poisson probabilities $p(0)$, $p(1)$, $p(2)$ and $p(3)$.

SOLUTION
$$p(r) = \frac{e^{-\lambda} \cdot \lambda^r}{r!}$$

$$\therefore \quad p(0) = \frac{e^{-2} \cdot 2^0}{0!} = \underline{0.135\,34}$$

$$p(1) = \frac{e^{-2} \cdot 2^1}{1!} = \underline{0.270\,67}$$

$$p(2) = \frac{e^{-2} \cdot 2^2}{2!} = \underline{0.270\,67}$$

$$p(3) = \frac{e^{-2} \cdot 2^3}{3!} = \underline{0.180\,45}$$

EXAMPLE 2.5.2 Show that the Poisson formula gives a good approximation of the Binomial probabilities when $n = 100$ and $p = 0.01$.

SOLUTION We have $n = 100$, $p = 0.01$, so $\lambda = np = 100(0.01) = 1$. Using the Binomial formula we have:

$$p(0) = {}^{100}C_0 (0.01)^0 (0.99)^{100} = 0.366\,03$$
$$p(1) = {}^{100}C_1 (0.01)^1 (0.99)^{99} = 0.369\,73$$
$$p(2) = {}^{100}C_2 (0.01)^2 (0.99)^{98} = 0.184\,87$$
$$p(3) = {}^{100}C_3 (0.01)^3 (0.99)^{97} = 0.061\,00 \quad \text{etc.}$$

Now using the Poisson formula we have:

$$p(0) = \frac{e^{-1} \cdot 1^0}{0!} = 0.367\,88$$

$$p(1) = \frac{e^{-1} \cdot 1^1}{1!} = 0.367\,88$$

$$p(2) = \frac{e^{-1} \cdot 1^2}{2!} = 0.183\,94$$

$$p(3) = \frac{e^{-1} \cdot 1^3}{3!} = 0.061\,31 \quad \text{etc.}$$

Since n is large and p small, the Binomial and Poisson probabilities are in close agreement: in this example they agree to two or three decimal places.

PROBABILITY

EXAMPLE 2.5.3 An assembly line produces approximately 2% defective items. In a batch of 140 items find the probability of obtaining:
(a) only two defective,
(b) less than two defective.

SOLUTION We have $n = 140$ and $p = 0.02$. Since n is large and p is small we can use the Poisson distribution to approximate these Binomial probabilities. Now $\lambda = np = 140(0.02) = 2.8$.

(a) $$p(2) = \frac{e^{-2.8}(2.8)^2}{2!} = \underline{0.238\,38}$$

(b) $$p(\text{less than } 2) = p(0) + p(1)$$

where $$p(0) = \frac{e^{-2.8}(2.8)^0}{0!} = 0.060\,81$$

and $$p(1) = \frac{e^{-2.8}(2.8)^1}{1!} = 0.170\,27$$

$\therefore \quad p(\text{less than } 2) = 0.060\,81 + 0.170\,27$

$$= \underline{0.231\,08}$$

EXAMPLE 2.5.4 Serious accidents occur at random in a particular manufacturing industry at the rate of 1.5 per week.

Find the probability of less than two accidents occurring during:
(a) a given week
(b) a four-week period.

SOLUTION (a) $\lambda = 1.5$

$$p(\text{less than } 2) = p(0) + p(1)$$

where $$p(0) = \frac{e^{-1.5}(1.5)^0}{0!} = 0.223\,13$$

and $$p(1) = \frac{e^{-1.5}(1.5)^1}{1!} = 0.334\,70$$

$\therefore \quad p(\text{less than } 2) = 0.223\,13 + 0.334\,70$

$$= \underline{0.557\,83}$$

(b) In four weeks $\lambda = 1.5 \times 4 = 6$.

$\therefore \quad p(0) = \frac{e^{-6}(6)^0}{0!} = 0.002\,48$

and $\quad p(1) = \frac{e^{-6}(6)^1}{1!} = 0.014\,87$

$\therefore \quad p(\text{less than } 2) = 0.002\,48 + 0.014\,87$

$$= \underline{0.017\,35}$$

EXERCISES ON SECTIONS 2.3 TO 2.5

1. Evaluate the following number of combinations: (a) $^5C_2$, (b) $^7C_6$, (c) $^6C_0$, (d) $^{10}C_{10}$.

2. If $n = 5$ and $p = 0.3$, use the Binomial formula to obtain the following probabilities: (a) $p(0)$, (b) $p(2)$, (c) p(less than 3).

3. If $\lambda = 2.5$, use the Poisson formula to obtain the following probabilities: (a) $p(0)$, (b) p(less than 2), (c) p(more than 2).

4. An assembly line produces approximately 10% defective items. In a sample of six items find the probability of obtaining:

    (a) no defectives

    (b) only one defective

    (c) more than one defective.

5. 25% of job applicants are rejected through lack of qualifications. In a group of four applicants find the probability that:

    (a) they all have suitable qualifications

    (b) more than two are under-qualified.

    State the mean and standard deviation of the number rejected from such a group.

6. It is assumed that approximately 3% of employees will resign during the next year. In a section containing 40 employees, find the chance that during the year:

    (a) no employees leave

    (b) more than two employees leave.

7. Customers arrive at a counter at random at an average rate of 0.6 per minute. Find the probability that:

    (a) no customers arrive during a 1 min period

    (b) at least two customers arrive during a 1 min period,

    (c) less than two customers arrive during a 10 min period.

2.6 NORMAL DISTRIBUTION

Consider a Normal distribution with mean $= \mu$ and standard deviation $= \sigma$ as shown below.

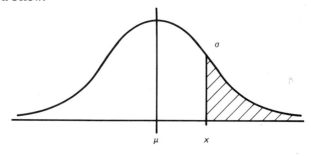

PROBABILITY 43

The area shaded can be obtained from tables of the Normal distribution by finding the standardised unit:

$$u = \frac{x - \mu}{\sigma}$$

[*Note*: The total area under the Normal curve = 1.]

EXAMPLE 2.6.1 Given a Normal distribution with mean = 30 and standard deviation = 5, find the areas under this Normal curve:

(a) above 40
(b) below 33
(c) above 22
(d) between 25 and 42.

SOLUTION (a)

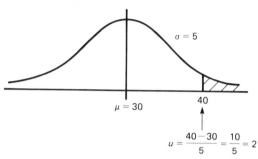

$$u = \frac{40 - 30}{5} = \frac{10}{5} = 2$$

From the Normal tables, area shaded = <u>0.022 75</u>

(b)

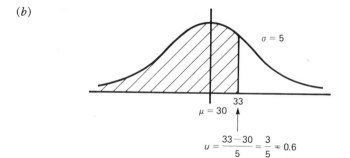

$$u = \frac{33 - 30}{5} = \frac{3}{5} = 0.6$$

Area shaded = $1 - 0.274\,25$ = <u>0.725 75</u>

(c)

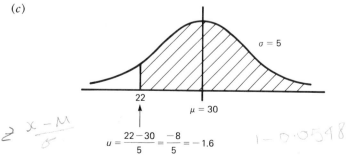

$$u = \frac{22 - 30}{5} = \frac{-8}{5} = -1.6$$

Area shaded = <u>0.945 20</u>

(d)

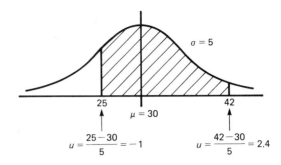

Area shaded = area above 25 − area above 42

= 0.841 34 − 0.008 20

= 0.833 14

EXAMPLE 2.6.2 The wages of blue-collar workers in a large company are Normally distributed with a mean of £110 per week and a standard deviation of £15 per week. Find the probability of a worker, selected at random, earning:

(a) over £130 per week,

(b) between £100 and £140 per week.

SOLUTION (a)

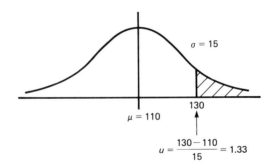

Required probability = area shaded = 0.091 76

(b)

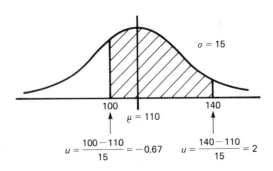

Required probability = area shaded = 0.748 57 − 0.022 75

= 0.725 82

EXAMPLE 2.6.3 The rateable value of housing in a given area is approximately Normally distributed with a mean of £310 and a standard deviation of £52. In a random selection of 600 houses, how many would you estimate to have rateable values of:

(a) less than £250

(b) more than £400?

SOLUTION (a)

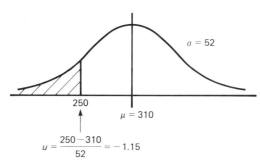

$$u = \frac{250-310}{52} = -1.15$$

Area shaded $= 1 - 0.874\,93$

$\qquad\qquad\quad = 0.125\,07$

∴ Expected number from 600 $= 600(0.125\,07)$

$\qquad\qquad\qquad\qquad\qquad\quad = 75.042$

∴ Approximately 75 houses will have rateable values of less than £250.

(b)

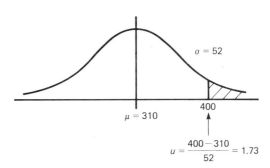

$$u = \frac{400-310}{52} = 1.73$$

Area shaded $= 0.041\,82$

∴ Expected number from 600 $= 600(0.041\,82)$

$\qquad\qquad\qquad\qquad\qquad\quad = 25.092$

∴ Approximately 25 houses will have rateable values above £400.

EXAMPLE 2.6.4 In an assessment of job performance the marks awarded are Normally distributed with a mean of 55 and a standard deviation of 11.

(a) In a group of 300 employees, how many would you expect to obtain over 75 marks?

(b) From past performances it can be seen that approximately 9% of employees obtain unsatisfactory gradings. What is the minimum 'satisfactory' grade?

SOLUTION (a)

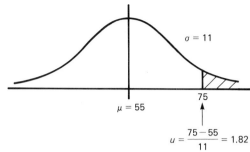

Area shaded $= 0.034\,38$

∴ In a group of 300 employees: $300(0.034\,38) = 10.314$

∴ Approximately 10 employees obtain over 75 marks.

(b)

M is the minimum satisfactory grade. From tables

$$\frac{M-55}{11} = -1.34$$

∴ $M - 55 = 11(-1.34) = -14.74$

∴ $M = 40.26$

∴ Minimum grade $= 40$ marks

EXAMPLE 2.6.5 The weights of packets of cereal from the Superbrek Company are Normally distributed with a mean of 505 g and a standard deviation of 8.2 g. Find the 95% confidence limits for the weight of such packets.

SOLUTION

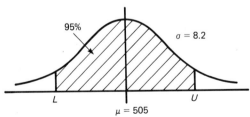

L and U are the 95% confidence limits. These limits are found by the formula $\mu \pm 1.96\sigma$. So

$$505 \pm 1.96(8.2) = 505 \pm 16.072$$

$$= 489 \text{ to } 521$$

∴ 95% of packets will weigh between 489 and 521 g.

EXERCISES ON SECTION 2.6

1. Given a Normal distribution with mean = 60 and standard deviation = 10, find the areas under the Normal curve:
   (a) over 70
   (b) under 62
   (c) over 54
   (d) between 62 and 72
   (e) between 55 and 80.

2. A population is Normal with $\mu = 35$ and $\sigma = 8$.
   (a) If one item is taken at random from this population find the probability that it is:
       (i) greater than 50
       (ii) less than 27
       (iii) between 30 and 45.
   (b) Find the 95% confidence limits for this item.

3. Hourly wage rates for unskilled workers in a particular nationwide industry are Normally distributed with a mean of £2.00 and a standard deviation of 35p.
   (a) Find the probability that an employee selected at random will earn a basic rate of between £1.90 and £2.20 per hour.
   (b) In a group of 400 unskilled employees how many would you expect to earn more than £2.50 per hour?
   (c) Approximately 20% earn less than the recommended minimum basic rate. What is this minimum rate?

4. A machine produces resistors with an average resistance of 10 ohms and a standard deviation of 0.3 ohms. The resistances are known to conform to the Normal distribution.
   (a) Find the probability that a resistor taken at random will have a resistance greater than 10.5 ohms.
   (b) 2% of resistors are over the accepted tolerance limit for such an item. Estimate this upper limit.
   (c) In a batch of 2000 resistors how many would you expect to have resistances under 9.3 ohms?

5. The total weekly sales from a company is approximately Normally distributed with an average of £80 000 and a standard deviation of £14 000.
   (a) In a week chosen at random, find the chance that total sales will not reach £50 000.
   (b) Over a full year (52 weeks!) estimate the number of weeks with sales exceeding £95 000.

PAST QUESTIONS 2

1. A manufacturer of portable generators and compressors obtains his 65 ml capacity engines from an outside supplier. Engines selected at random are tested for power output under specific conditions. A batch of 36 of these engines produced the following results, in kilowatts:

0.59	0.61	0.59	0.60	0.60	0.61	0.60	0.61	0.58
0.62	0.60	0.60	0.57	0.62	0.62	0.64	0.59	0.61
0.59	0.63	0.60	0.61	0.60	0.60	0.61	0.63	0.60
0.62	0.61	0.58	0.63	0.61	0.64	0.59	0.61	0.60

   (a) Form the data into a suitable frequency distribution, and hence calculate the arithmetic mean and standard deviation.

   (b) Assuming the distribution to be close to Normal, obtain the 95% confidence interval for the power output of all such engines obtained from the same supplier. [IIM]

2. In an acceptance sampling scheme, a random sample of size 100 items is taken from a batch and inspected, and if it contains 3 or fewer defectives the batch is accepted without further inspection, and is said to have passed the sampling scheme. If the sample contains 4 or more defectives then all the remaining items in the batch are inspected and the batch is said to have failed the scheme.

   Required:

   (a) By using the Poisson distribution to calculate the probability of finding 0, 1, 2 and 3 defectives, determine the proportion of batches containing 2% defective items that will pass the acceptance sampling scheme.

   (b) Show that the expected number of defectives found by the scheme in a batch containing 2% defectives that passes the scheme is approximately 1.6. By considering, in addition, those batches that fail the scheme, determine the expected number of defectives in batches of size 2000 containing 2% defectives. [ACA]

3. (a) An item produced by a company is susceptible to two types of defect, A and B. The probability that an item has defect A is $\frac{1}{6}$. The probability that it has defect B is $\frac{1}{8}$, independent of whether it has defect A.

   Required:

   (i) calculate the probability that an item has:
   1. both A and B defects,
   2. one defect only, A or B,
   3. no defect.

   (ii) What simple relationship is there between your answers to part (i)?

   (b) Suppose now, in addition to the above defects, the item is susceptible to a third type of defect C. The probability that an item contains C depends on whether it has the other defects. If it has neither A nor B there is a probability of $\frac{1}{10}$ that it has C. If it has one of A or B the probability of having C is $\frac{2}{10}$, and if it has both A and B the probability of having C is $\frac{3}{10}$.

Required:
(i) Show that the probability that an item has
 1. none of the three defects is $\frac{315}{480}$,
 2. one of the three defects is $\frac{131}{480}$.
(ii) If items with one defect can be repaired at a cost of £10 but those with two or more defects are scrapped at a cost of £30, determine the total cost of repair and scrapping associated with the production of 480 items.   [ACA]

4. (a) In a large batch of finished units there are known to be 10% defectives. If a random sample of five units is selected from the batch, calculate the probability that the sample will contain:
 (i) no defectives,
 (ii) one defective,
 (iii) more than three defectives.
(b) In a particular organisation the number of hours lost through absenteeism each week is Normally distributed with a mean of 180 hours and a standard deviation of 15 hours. If each week can be treated as being independent, what is the probability of more than:
 (i) 720 total hours being lost in a four-week period,
 (ii) 800 total hours being lost in a four-week period,
 (iii) 210 hours being lost in each of the four weeks?   [IMS]

5. Manufactured items are sold in boxes which are stated to contain a weight of at least 40 ounces. The actual weight in a box varies, being approximately Normally distributed with mean 41.2 ounces and standard deviation 0.8 ounces.

Required:
(a) Calculate the proportion of boxes whose weight is between 40 ounces and 42 ounces.
(b) Calculate the weight below which 20% of boxes fall.
(c) All boxes containing less than 40 ounces are scrapped at a cost of £1 per box. Calculate the scrapping cost associated with the sale of 100 boxes.
(d) To what mean weight should the box contents be adjusted, the standard deviation remaining unchanged, if only 1% of boxes are to be scrapped?
           [ACA]

6. The lifetimes of a certain kind of battery have a mean of 300 hours and a standard deviation of 35 hours. Assuming that the distribution of the lifetimes, which are measured to the nearest hour, can be approximated closely with a Normal curve, ascertain:
(a) the percentage of the batteries which have a lifetime of more than 320 hours,
(b) the value above which the best 30 per cent of the batteries will lie,
(c) the proportion of the batteries that have a lifetime from 250 to 350 hours inclusive,

(d) the types of data you would expect to conform to the Normal distribution,

(e) how you would verify whether data conformed to the Normal distribution or not. [CIT]

7. (a) Bus-Hire Limited has two coaches which it hires out for local use by the day. The number of demands for a coach on each day is distributed as a Poisson distribution, with a mean of two demands.
   (i) On what proportion of days is neither coach used?
   (ii) On what proportion of days is at least one demand refused?
   (iii) If each coach is used an equal amount, on what proportion of days is *one* particular coach not in use?

(b) Steel rods are manufactured to a specification of 20 cm length and are acceptable only if they are within the limits of 19.9 cm and 20.1 cm. If the lengths are normally distributed, with mean 20.02 and standard deviation 0.05 cm, find the percentage of rods which will be rejected as:
   (i) undersize,
   (ii) oversize. [ICMA]

8. Two types of candidate, A and B, are examined in a single examination paper. The marks for type A are Normally distributed with mean 52 and standard deviation 10. The marks for type B are Normally distributed with mean 62 and standard deviation 12. The pass mark is 50.

Required:

(a) Calculate the proportion of type A candidates who:
   (i) pass the examination,
   (ii) obtain between 50 and 60 marks.

(b) Calculate the mark exceeded by the top 20% of type A candidates.

(c) If two-thirds of candidates are type A, one-third are type B, calculate the overall percentage of candidates who pass the examination.

(d) Calculate the overall mean mark in the examination. [ACA]

9. (a) Explain what is meant by mutually exclusive events and independent events in the context of probability.

(b) A component passes through five operations in sequence before it is completed. The following information on reject rates is available.

	Number of units introduced	3200
	1st operation	136
	2nd operation	126
Rejects in	3rd operation	54
	4th operation	48
	5th operation	12

   (i) Estimate the probability of a component reaching the third stage of operation.
   (ii) Estimate the probability that both components in a sample of two will be completed.
   (iii) If a further 600 finished components are required, estimate how many components should be introduced. [IIM]

10. (a) The independent probabilities that the three sections of a costing department will encounter a computer error are respectively 0.1, 0.2 and 0.3 each week. Calculate the probability that there will be:
    (i) at least one computer error,
    (ii) one and only one computer error,
    encountered by the costing department next week.

    (b) Experience has shown that, on the average, 2% of an airline's flights suffer a minor equipment failure in an aircraft. Use the Poisson distribution to estimate the probability that the number of minor equipment failures in the next 50 flights will be:
    (i) zero,
    (ii) at least two. [ICMA]

11. (a) Describe the main features of the Normal distribution, indicate why the distribution is of fundamental importance in applications of statistics and give *three* examples of random variables which follow this distribution.

    (b) Breakdowns of an automatic welding machine occur as events in a Poisson process at an average rate of 9 per week. What is the probability that the number of breakdowns during any specified four-week period:
    (i) is less than 27,
    (ii) lies between 42 and 45? [IPM]

12. (a) A company produces batteries whose lifetimes are Normally distributed with a mean of 100 hours. It is known that 90% of the batteries last at least 40 hours.
    (i) Estimate the standard deviation lifetime,
    (ii) What percentage of batteries will not last 70 hours?

    (b) A company mass produces electronic calculators. From past experience it knows that 90% of the calculators will be in working order and 10% will be faulty if the production process is working satisfactorily. An inspector randomly selects 5 calculators from the production line every hour and carries out a rigorous check.
    (i) What is the probability that a random sample of 5 will contain at least 3 defective calculators?
    (ii) A sample of 5 calculators is found to contain 3 defectives; do you consider the production process to be working satisfactorily? [ICMA]

13. The average number of canteen meals per year purchased by employees was 400 and the standard deviation was 100. Assuming that meal purchases were Normally distributed, what is the probability that employees bought:

    (a) between 250 and 500 meals,
    (b) less than 250 meals,
    (c) between 500 and 600 meals,
    (d) more than 500 meals?

    (Use the formula:

    $$z = \frac{x - \mu}{\sigma}$$

where z = number of standard deviations from the mean, x = any specified value of the variable, μ = the mean of the distribution, and σ = the standard deviation of the distribution.) [ITD]

14. A group of five trainees are to be given a practical test. The probability that any one of the trainees passes is estimated at 0.9. Calculate the probability of the following outcomes of the test:

(a) all five trainees pass,

(b) one passes and the other four fail,

(c) more than three pass.

Name the probability distribution which could fit the above situation and calculate the expected number of passes and also the standard deviation of the distribution. [IMS]

15. Assume that the population mean for weekly wages is £110 and the population standard deviation is £12. What is the probability of a person earning £134 or above? Less than £80? Between £104 and £122? [ITD]

16. (a) In what circumstances should the probabilities of two or more simple events be added together and how would the result be altered if the events are not mutually exclusive? When are the probabilities multiplied together and how is the result affected if the events are not independent?

(b) A firm submits tenders for four separate contracts and the probabilities of acceptance are 0.3, 0.6, 0.5 and 0.4 respectively. If the chances of success are independent what is the probability that:
  (i) all,
  (ii) none,
  (iii) only one,
  (iv) at least two,
  are successful? [RICS]

3 SAMPLING DISTRIBUTIONS

3.1 DISTRIBUTION OF SAMPLE MEANS

Consider a population with mean $= \mu$ and standard deviation $= \sigma$. Samples of size n are taken from this population and the sample means $\bar{x}$ are found. The distribution of sample means has mean

$$\mu_{\bar{x}} = \mu$$

and standard deviation (standard error)

$$\sigma_{\bar{x}} = \frac{\sigma}{\sqrt{n}}$$

If the population is Normally distributed, or if the sample size is 'large' (i.e. $n > 30$), then the distribution of sample means is approximately Normal.

EXAMPLE 3.1.1 A Normal population has a mean $= 200$ and standard deviation $= 50$. Find the probability that a sample of 25 values has a mean greater than 215.

SOLUTION We have $\mu = 200$ and $\sigma = 50$. The distribution of sample means has mean

$$\mu_{\bar{x}} = \mu = 200$$

and standard error

$$\sigma_{\bar{x}} = \frac{\sigma}{\sqrt{n}} = \frac{50}{\sqrt{25}} = \frac{50}{5} = 10$$

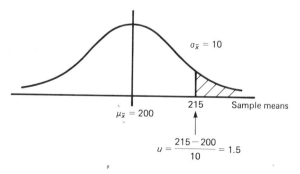

Therefore the probability that a sample mean is greater than 215 is equal to the area shaded $= \underline{0.066\,81}$ from tables.

EXAMPLE 3.1.2 The daily output from a production line has a mean of 6000 units with a standard deviation of 400 units. What is the chance that during the next 100 days the average output will be under 5900 units per day?

SOLUTION We have $\mu = 6000$ and $\sigma = 400$. The distribution of sample means is approximately Normal (since n is large) with

$$\mu_{\bar{x}} = \mu = 6000$$

and

$$\sigma_{\bar{x}} = \frac{\sigma}{\sqrt{n}} = \frac{400}{\sqrt{100}} = \frac{400}{10} = 40$$

$$u = \frac{5900 - 6000}{40} = \frac{-100}{40} = -2.5$$

$p(\text{sample mean} < 5900) = \text{area shaded}$

$\phantom{p(\text{sample mean} < 5900)} = \underline{0.006\,21} \quad \text{from tables}$

EXAMPLE 3.1.3 Find the 95% confidence limits for the average daily output over 100 days given in Example 3.1.2.

SOLUTION We have $\mu = 6000$ and $\sigma = 400$. Also $\mu_{\bar{x}} = 6000$, $\sigma_{\bar{x}} = 40$. The 95% confidence limits for a sample mean are:

$$\mu_{\bar{x}} \pm 1.96\sigma_{\bar{x}} = 6000 \pm 1.96(40)$$

$$= 6000 \pm 78.4$$

$$= 5921.6 \quad \text{to} \quad 6078.4$$

We are 95% sure that the average output over 100 days will be between $\underline{5921.6 \text{ and } 6078.4}$ units per day.

3.2 SIGNIFICANCE OF A SAMPLE MEAN

To investigate the significance of a sample mean we evaluate:

$$u = \frac{\bar{x} - \mu_{\bar{x}}}{\sigma_{\bar{x}}}$$

i.e.

$$u = \frac{\bar{x} - \mu}{\sigma/\sqrt{n}}$$

SAMPLING DISTRIBUTIONS 55

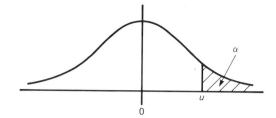

A selection of 'significant' values of u together with the significance level α are given below.

α	5%	$2\frac{1}{2}$%	1%	$\frac{1}{2}$%
u	1.64	1.96	2.33	2.58

EXAMPLE 3.2.1 Consider a Normal population with a standard deviation = 20. A random sample of 16 items is found to have a mean of 112. Test the assumption at the 5% significance level that the population has a mean of 100.

SOLUTION Assumption: population mean = 100. Alternative: population mean ≠ 100. We write H_0: $\mu = 100$, H_1: $\mu \neq 100$.

We are given that $\sigma = 20$, $n = 16$ and $\bar{x} = 112$. Now

$$u = \frac{\bar{x} - \mu}{\sigma/\sqrt{n}} = \frac{112 - 100}{20/\sqrt{16}}$$

$$\therefore \quad u = \frac{12}{20/4} = \frac{12}{5} = \underline{2.4}$$

A 'significant' value of u at the 5% level is 1.96, i.e. the 95% confidence limits for u are -1.96 to $+1.96$ (see diagram).

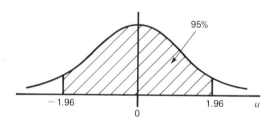

Therefore, our value of u is significant (i.e. it is outside the confidence limits). We reject H_0. We thus accept that the population mean is not equal to 100.

EXAMPLE 3.2.2 A random sample of 10 family cars is found to have an average retail price of £6620. Assuming that car prices are Normally distributed with a standard deviation of £1100, test the assumption (at the 5% level) that the average price of a family car is: (a) £6000 and (b) more than £6000.

SOLUTION We are given that $\sigma = 1100$, $\bar{x} = 6620$, $n = 10$.

(a) H_0: $\mu = 6000$, H_1: $\mu \neq 6000$ (this is a two-tailed test). We have

$$u = \frac{\bar{x} - \mu}{\sigma/\sqrt{n}} = \frac{6620 - 6000}{1100/\sqrt{10}}$$

$$= \frac{620}{1100/3.1623} = \underline{1.782}$$

Now a significant value of u at the 5% level is 1.96. Therefore, our value of u (= 1.782) is not significant. We accept the assumption H_0. Thus the sample shows that the average price of a family car is not significantly different from £6000.

(b) H_0: $\mu = 6000$, H_1: $\mu > 6000$ (this is a one-tailed test). We have

$$u = \frac{\bar{x} - \mu}{\sigma/\sqrt{n}} = \frac{6620 - 6000}{1100/\sqrt{10}} = \underline{1.782}$$

Now a significant value of u at the 5% level (one tail) is 1.64.

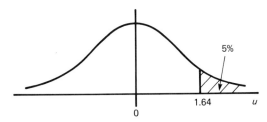

Therefore, our value of u (= 1.782) is significant. We reject H_0. We thus accept the assumption that the average family car costs more than £6000.

EXERCISES ON SECTIONS 3.1 AND 3.2

1. Given a population with a mean = 300 and standard deviation = 30. Samples of 36 values are taken and their means are found. Find the mean and standard deviation of the distribution of sample means.

2. A population has a mean = 74 with a standard deviation = 8. Find the probability that a random sample of 100 items has a mean: (a) greater than 75, and (b) between 73.5 and 74.5.

3. The values of orders received by a company are Normally distributed with a mean of £3400 and a standard deviation of £1150. In a batch of 25 orders find the probability that the average value is:
   (a) in excess of £4000
   (b) below £3000 and
   (c) between £3500 and £3700.

4. The average income tax allowance for employees in a company is £3900 per annum, with a standard deviation of £750.
   (a) Find the probability that a group of 60 employees, selected at random, will have an average income tax allowance of (i) over £4000 per annum, and (ii) between £3600 and £3800 per annum.

(b) Find the 95% confidence limits for the average tax allowance in such a group.

5. A Normal population has a standard deviation of 30. A random sample of 40 items is found to have a mean of 260. Using the 5% significance level examine the assumption that the population has a mean of 250.

6. A random sample of 80 houses is found to have an average price of £33 000. At the 1% level test the assumption that the average house price exceeds £31 000. (It can be assumed that house prices are Normally distributed with a standard deviation of £7000.)

7. Weekly wages in a company are Normally distributed with a standard deviation of £21. A sample of 25 employees is found to have a mean wage of £168 per week. Using the 5% level of significance would you conclude that the average wage in this company is significantly higher than £160 per week?

3.3 THE t DISTRIBUTION

Given a Normal population with mean $= \mu$ and standard deviation unknown, we can examine the 'significance' of a sample mean by using the t-test:

$$t = \frac{\bar{x} - \mu}{s/\sqrt{n-1}}$$

where s = sample standard deviation, with the number of degrees of freedom $\nu = n - 1$.

EXAMPLE 3.3.1 Find the value of t exceeded by only 5% in the t distribution given that (a) $\nu = 2$, and (b) $\nu = 10$.

SOLUTION

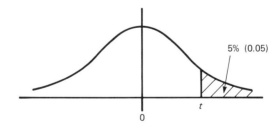

We require the value of t shown in the above diagram. From tables of the t distribution we have:

(a)   when $\nu = 2$, $t =$ <u>2.920</u>

(b)   when $\nu = 10$, $t =$ <u>1.812</u>

EXAMPLE 3.3.2 Find a 'significant' value of t at the 1% level with (a) $\nu = 4$, and (b) $\nu = 8$.

SOLUTION We require the value of t as illustrated below.

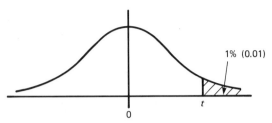

From the t distribution tables we have:

(a)   when $\nu = 4$, $t =$ <u>3.747</u>

(b)   when $\nu = 8$, $t =$ <u>2.896</u>

EXAMPLE 3.3.3 A random sample of 10 items is obtained from a Normal population and is found to have a mean = 40 with a standard deviation = 5. Test the assumption, at the 5% significance level, that the population mean is 38.

SOLUTION We are given $\bar{x} = 40$, $s = 5$, $n = 10$. We write $H_0: \mu = 38$, $H_1: \mu \neq 38$ (two-tailed test). Now

$$t = \frac{\bar{x} - \mu}{s/\sqrt{n-1}} = \frac{40 - 38}{5/\sqrt{10-1}}$$

$$= \frac{2}{5/\sqrt{9}} = 1.2$$

Now, a significant value of t at the 5% level (two tails) with $\nu = n - 1 = 10 - 1 = 9$ is 2.262. Therefore, our value of t ($= 1.2$) is not significant. We accept H_0. We thus conclude that the population mean does not differ significantly from 38.

EXAMPLE 3.3.4 The weekly turnover of a retail company is Normally distributed with an average of £42 000 per week.

Following an advertising campaign a seven-week period produced an average turnover of £49 500 per week with a standard deviation of £4500. At the 1% level test whether there has been a significant increase in the turnover.

SOLUTION We are given $\bar{x} = 49\,500$, $s = 4500$, $n = 7$. We write $H_0: \mu = 42\,000$, $H_1: \mu > 42\,000$ (one tail). Now

$$t = \frac{\bar{x} - \mu}{s/\sqrt{n-1}} = \frac{49\,500 - 42\,000}{4500/\sqrt{7-1}}$$

$$= \frac{7500}{4500/\sqrt{6}} = 4.082$$

Now a significant value of t at the 1% level (one tail) with $\nu = n - 1 = 7 - 1 = 6$ is 3.143. Therefore, our value of t ($= 4.082$) is significant. We reject H_0. It thus seems that there has been a significant increase in the weekly turnover.

SAMPLING DISTRIBUTIONS 59

EXAMPLE 3.3.5 The amount of monthly income tax paid by employees is approximately Normally distributed. A random sample of 30 employees paid an average of £105 per month in income tax, with a standard deviation of £48 per month.

At the 5% significance level test the assumption that the average amount of income tax paid is greater than £80 per month per employee.

SOLUTION We are given $\bar{x} = 105$, $s = 48$, $n = 30$. We write $H_0: \mu = 80$, $H_1: \mu > 80$ (one tail). Now

$$t = \frac{\bar{x} - \mu}{s/\sqrt{n-1}} = \frac{105 - 80}{48/\sqrt{29}}$$

$$= 2.805$$

From tables, a significant value of t at the 5% level with $\nu = 29$ is 1.699. Therefore, our value of $t (= 2.805)$ is significant. We can reject H_0. We thus accept the assumption that the average income tax paid by employees in the company is significantly greater than £80 per month.

EXAMPLE 3.3.6 An assessment test is given to all prospective employees in a company. The test scores are known to be Normally distributed. A random sample of five participants obtained the following results: 48, 56, 62, 65, 74.

Test the assumption that the mean test score is 55 using the 5% significance level.

SOLUTION $H_0: \mu = 55$, $H_1: \mu \neq 55$ (two tails). We have x: 48, 56, 62, 65, 74. Now

$$\bar{x} = \frac{\Sigma x}{n} = \frac{305}{5} = 61$$

and

$$s = \sqrt{\frac{\Sigma(x - \bar{x})^2}{n}} = \sqrt{\frac{380}{5}} = \sqrt{76} = 8.7178$$

$$\therefore \quad t = \frac{\bar{x} - \mu}{s/\sqrt{n-1}} = \frac{61 - 55}{8.7178/\sqrt{4}} = 1.376$$

From tables, a significant value of t at the 5% level with $\nu = 4$ is 2.776. Therefore, our value of $t (= 1.376)$ is not significant. We can accept H_0. We thus conclude that the average test score is not significantly different from 55.

3.4 DIFFERENCE BETWEEN TWO SAMPLE MEANS

Two samples are taken from approximately Normal populations with means μ_1 and μ_2 and a common variance, giving the following results:

sample 1: mean = $\bar{x}_1$, standard deviation = s_1, size = n_1

sample 2: mean = $\bar{x}_2$, standard deviation = s_2, size = n_2

To investigate whether μ_1 and μ_2 are significantly different we use the t-test

with:
$$t = \frac{|\bar{x}_1 - \bar{x}_2|}{\hat{\sigma}\sqrt{1/n_1 + 1/n_2}}$$

where

$$\hat{\sigma}^2 = \frac{n_1 s_1^2 + n_2 s_2^2}{n_1 + n_2 - 2}$$

with the number of degrees of freedom $\nu = n_1 + n_2 - 2$.

EXAMPLE 3.4.1 Two random samples are taken from a population. The first sample of 20 values has a standard deviation = 10. The second sample of 30 values has a standard deviation = 12. Estimate the population standard deviation.

SOLUTION We have $n_1 = 20$, $s_1 = 10$; $n_2 = 30$, $s_2 = 12$. The 'best' estimate of the population variance is:

$$\hat{\sigma}^2 = \frac{n_1 s_1^2 + n_2 s_2^2}{n_1 + n_2 - 2} = \frac{20 \times 10^2 + 30 \times 12^2}{20 + 30 - 2}$$

$$= \frac{6320}{48} = 131.667$$

So the best estimate of the population standard deviation is:

$$\hat{\sigma} = \sqrt{131.667} = 11.475$$

EXAMPLE 3.4.2 Use the t-test to investigate whether there is a significant difference between the two sample means given below:

sample 1: mean = 43, standard deviation = 6, size = 15

sample 2: mean = 38, standard deviation = 10, size = 30

(Use the 5% significance level.)

SOLUTION $H_0: \mu_1 = \mu_2$, $H_1: \mu_1 \ne \mu_2$ (two-tailed test). We are given $\bar{x}_1 = 43$, $s_1 = 6$, $n_1 = 15$ and $\bar{x}_2 = 38$, $s_2 = 10$, $n_2 = 30$. Now

$$\hat{\sigma}^2 = \frac{n_1 s_1^2 + n_2 s_2^2}{n_1 + n_2 - 2} = \frac{15 \times 6^2 + 30 \times 10^2}{15 + 30 - 2}$$

$$= \frac{540 + 3000}{43} = 82.3256$$

$\therefore \quad \hat{\sigma} = \sqrt{82.3256} = 9.0733$

$\therefore \quad t = \frac{|\bar{x}_1 - \bar{x}_2|}{\hat{\sigma}\sqrt{1/n_1 + 1/n_2}} = \frac{|43 - 38|}{9.0733\sqrt{1/15 + 1/30}}$

$\therefore \quad t = \frac{5}{2.8692} = 1.743$

Now a significant value of t at the 5% level with $\nu = n_1 + n_2 - 2 = 43$ is approximately 2.02. Therefore, our value of $t (= 1.743)$ is not significant. We accept H_0. There is thus no significant difference between the two sample means.

EXAMPLE 3.4.3 Random samples of employees are taken from two companies in order to compare earnings. A sample of 40 employees from Company A is found to have a mean of £170 per week with a standard deviation of £24. A sample of 20 employees from Company B has a mean of £157 per week with a standard deviation of £17. Do these samples indicate a significant difference between the average earnings in the two companies at the 5% level?

SOLUTION We are given $\bar{x}_1 = 170$, $s_1 = 24$, $n_1 = 40$ and $\bar{x}_2 = 157$, $s_2 = 17$, $n_2 = 20$.
$H_0: \mu_1 = \mu_2$, $H_1: \mu_1 \neq \mu_2$ (two-tailed).

$$\hat{\sigma}^2 = \frac{n_1 s_1^2 + n_2 s_2^2}{n_1 + n_2 - 2} = \frac{40 \times 24^2 + 20 \times 17^2}{40 + 20 - 2}$$

$$= \frac{28\,820}{58} = 496.896$$

$\therefore \quad \hat{\sigma} = 22.291$

Now

$$t = \frac{\bar{x}_1 - \bar{x}_2}{\hat{\sigma}\sqrt{1/n_1 + 1/n_2}} = \frac{170 - 157}{22.291\sqrt{1/40 + 1/20}}$$

$$= \frac{13}{6.104\,64} = \underline{2.1295}$$

A significant value of t at the 5% level with $\nu = 40 + 20 - 2 = 58$ is approximately 2.00. Therefore our value of $t (= 2.1295)$ is significant. We reject H_0. We thus conclude that there is a significant difference between the average earnings in the two companies.

EXAMPLE 3.4.4 In a survey of house prices in two areas the following information has been obtained.

Area	No. of houses sampled	Average price (£)	Standard deviation (£)
X	68	32 000	6000
Y	54	29 000	5000

Are house prices in area X significantly more than those in area Y? (Use the 5% level of significance.)

SOLUTION We have $\bar{x}_1 = 32\,000$, $s_1 = 6000$, $n_1 = 68$, $\bar{x}_2 = 29\,000$, $s_2 = 5000$, $n_2 = 54$.
$H_0: \mu_1 = \mu_2$, $H_1: \mu_1 > \mu_2$ (one-tailed test).

$$\hat{\sigma}^2 = \frac{n_1 s_1^2 + n_2 s_2^2}{n_1 + n_2 - 2} = \frac{68(6000)^2 + 54(5000)^2}{68 + 54 - 2}$$

$$= 31\,650\,000$$

$$\therefore \quad \hat{\sigma} = 5625.833$$

Now

$$t = \frac{\bar{x}_1 - \bar{x}_2}{\hat{\sigma}\sqrt{1/n_1 + 1/n_2}} = \frac{32\,000 - 29\,000}{5625.833\sqrt{1/68 + 1/54}}$$

$$= \frac{3000}{1025.45} = \underline{2.926}$$

Now a significant value of t at the 5% level (one tail) with $\nu = n_1 + n_2 - 2 = 68 + 54 - 2 = 120$ is 1.658. Therefore, our value of $t(= 2.926)$ is significant. We reject H_0. The evidence thus suggests that houses in area X are more expensive than those in area Y.

3.5 THE PAIRED t-TEST

In many cases, when considering two samples of the same size n, it is possible to 'pair off' corresponding values in each sample. Given such a situation, in order to examine whether there is a significant difference between the sample means the *paired t-test* can be used with:

$$t = \frac{\bar{d}}{s_d/\sqrt{n-1}}$$

with $\nu = n - 1$, where $\bar{d}$ = mean of differences between pairings and s_d = standard deviation of differences.

EXAMPLE 3.5.1 Use the paired t-test to examine whether there is a significant difference between the two sets of figures (A and B) shown below. (Use the 5% level.)

	Sample no.					
	1	2	3	4	5	6
A	20	45	28	36	52	31
B	24	40	31	30	47	34

SOLUTION The differences between pairings are d: $-4, 5, -3, 6, 5, -3$.

$$\therefore \quad \bar{d} = \frac{-4 + 5 - 3 + 6 + 5 - 3}{6} = \frac{6}{6} = 1$$

and

$$s_d = \sqrt{\frac{(-4-1)^2 + (5-1)^2 + (-3-1)^2 + (6-1)^2 + (5-1)^2 + (-3-1)^2}{6}}$$

$$= \sqrt{\frac{25+16+16+25+16+16}{6}}$$

$$= \sqrt{19} = \underline{4.3589}$$

We assume that there is no difference between the sample means (i.e. $H_0: \mu_d = 0$, $H_1: \mu_d \neq 0$). Now

$$t = \frac{\bar{d}}{s_d/\sqrt{n-1}} = \frac{1}{4.3589/\sqrt{6-1}}$$

$$\therefore \quad t = \underline{0.513}$$

A significant value of t at the 5% level (two tails) with $\nu = 5$ is 2.571. Therefore, our value of $t(=0.513)$ is not significant. We thus conclude that there is no significant difference between the two samples A and B.

EXAMPLE 3.5.2 The scores for a sample of seven employees in two assessment tests are given below. Find whether the scores in Test I are significantly higher than those in Test II using the 1% level.

Employee	A	B	C	D	E	F	G
Test I (%)	48	51	54	58	60	66	73
Test II (%)	50	47	50	53	60	70	66

SOLUTION We assume that there is no difference between the two sets of scores ($H_0: \mu_d = 0$, $H_1: \mu_d > 0$). The differences are $d: -2, 4, 4, 5, 0, -4, 7$.

$$\therefore \quad \bar{d} = \frac{-2+4+4+5+0-4+7}{7} = \frac{14}{7} = 2$$

and

$$s_d = \sqrt{\frac{(-2-2)^2+(4-2)^2+(4-2)^2+(5-2)^2+(0-2)^2+(-4-2)^2+(7-2)^2}{7}}$$

$$= \sqrt{\frac{98}{7}} = \sqrt{14} = 3.74166$$

$$\therefore \quad t = \frac{\bar{d}}{s_d/\sqrt{n-1}} = \frac{2}{3.74166/\sqrt{6}} = \underline{1.309}$$

Now a significant value of t at the 1% level (one tail) with $\nu = 6$ is 3.143. Therefore, our value of $t(=1.309)$ is not significant. We thus conclude that the scores in Test I are not significantly higher than those in Test II.

EXAMPLE 3.5.3 The table below gives the prices of a random sample of 10 shares taken on two dates in a given year. Using the 5% significance level determine whether there has been a significant change in the value of the shares during the indicated period.

Share		1	2	3	4	5	6	7	8	9	10
Sale price (p)	1st Jan	85	263	165	342	154	95	305	127	69	28
	1st June	79	250	172	310	142	76	312	102	42	24

64 NOTES AND PROBLEMS IN STATISTICS

SOLUTION $H_0: \mu_d = 0$, $H_1: \mu_d \neq 0$. The differences are d: 6, 13, −7, 32, 12, 19, −7, 25, 27, 4.

$$\therefore \quad \bar{d} = 12.4 \quad \text{and} \quad s_d = 12.9012$$

$$t = \frac{\bar{d}}{s_d/\sqrt{n-1}} = \frac{12.4}{12.9012/\sqrt{10-1}}$$

$$= \underline{2.883}$$

Now a significant value of t at the 5% level (two tails) with $\nu = 9$ is 2.262. Therefore, our value of $t(= 2.883)$ is significant. We reject H_0. There is thus evidence to suggest that a significant change in share prices has occurred during the period 1st January to 1st June.

EXERCISES ON SECTIONS 3.3 TO 3.5

1. Find a 'significant' value of t at the 5% level when the number of degrees of freedom is (a) 2, (b) 5, and (c) 30.

2. A sample of 30 items from a Normal population is found to have a mean of 450 with a standard deviation of 55. At the 1% level test the assumption that the population has a mean of 420.

3. Using the 5% level of significance, test the assumption that a Normal population has a mean greater than 105 based on a sample of 65 items with a mean = 110 and a standard deviation = 12.

4. The monthly profit obtained by a company is approximately Normally distributed. During an eight-month period the mean profit is £25 500 per month with a standard deviation of £5400 per month. Test the assumption at the 5% level that the average profit in the company is significantly greater than £22 000 per month.

5. A random sample of six bank accounts showed balances equal to: £54, £640, £248, £320, £124, £396. Test the assumption that the mean bank balance is £240. (Use the 1% significance level.)

6. Two random samples are taken from a population giving the following results:

   sample 1: size = 6, standard deviation = 20
   sample 2: size = 16, standard deviation = 24

   Estimate the population standard deviation.

7. Using a 5% level of significance, investigate whether there is a signifcant difference between the two sample means shown below:

   sample A: mean = 52, standard deviation = 5, size = 20
   sample B: mean = 42, standard deviation = 4, size = 12

8. The following table shows the distribution of weekly wages of samples of employees from two companies.

Company	Sample size	Mean	Standard deviation
A	40	£180	£15
B	50	£173	£16

The Personnel Manager in Company A claims that wages in his company are significantly higher than those in Company B. Test this claim using the 1% level of significance.

9. Test whether there is a significant difference between the survival times of employees in two departments given the data obtained from random samples shown below.

Department	No. of employees sampled	Length of service (years)	
		Average	Standard deviation
Design	12	7.5	3.0
Sales	30	5.7	2.1

(Use the 5% significance level.)

10. Using a paired *t*-test examine whether there is any significant difference between the two samples shown below.

Sample A	4	6	12	10	11	7	3	9
Sample B	5	3	10	5	6	8	4	5

11. The table below gives the production figures in two successive weeks for a sample of assembly sections (A to I) in an electronics company. At the beginning of the second week a revised bonus scheme has been introduced in an attempt by the management to improve productivity. Comment on the effectiveness of this scheme.

	Units produced per hour								
	A	B	C	D	E	F	G	H	I
Week 1	46	38	52	48	45	50	41	36	45
Week 2	48	41	51	49	45	51	46	41	47

3.6 DISTRIBUTION OF PROPORTIONS

A population contains a proportion p of 'successes'. Random samples of size n are taken from this population. The proportions of 'successes' in the samples are distributed with a mean

$$\mu_{\hat{p}} = p$$

and standard deviation

$$\sigma_{\hat{p}} = \sqrt{\frac{p(1-p)}{n}}$$

If n is 'large' the sample proportions are approximately Normally distributed.

The 'significance' of a sample proportion $\hat{p}$ can be examined using the formula

$$u = \frac{\hat{p} - \mu_{\hat{p}}}{\sigma_{\hat{p}}}$$

Similarly, the difference between two sample proportions ($\hat{p}_1$ and $\hat{p}_2$) can be examined by

$$u = \frac{|\hat{p}_1 - \hat{p}_2|}{\sqrt{\hat{p}(1-\hat{p})(1/n_1 + 1/n_2)}}$$

where

$$\hat{p} = \frac{n_1 \hat{p}_1 + n_2 \hat{p}_2}{n_1 + n_2}$$

EXAMPLE 3.6.1 Given $n = 400$ and $p = 0.2$ find $\mu_{\hat{p}}$ and $\sigma_{\hat{p}}$.

SOLUTION Now $\mu_{\hat{p}} = p = 0.2$ and

$$\sigma_{\hat{p}} = \sqrt{\frac{p(1-p)}{n}} = \sqrt{\frac{0.2(1-0.2)}{400}}$$

$$= \sqrt{0.0004} = 0.02$$

$\therefore$ $\underline{\mu_{\hat{p}} = 0.2 \text{ and } \sigma_{\hat{p}} = 0.02}$

EXAMPLE 3.6.2 A manufacturing process produces approximately 10% defective items.

(a) Find the mean and standard deviation of the proportion of defectives obtained in samples of 200 items.

(b) Find the 95% confidence limits for the proportion of defectives in a sample of 200 items.

SOLUTION (a) We have $p = 0.1$ (i.e. 10%) and $n = 200$.

$\therefore$ Mean $= \mu_{\hat{p}} = p = \underline{0.1 \; (10\%)}$

Standard deviation $= \sigma_{\hat{p}} = \sqrt{\frac{p(1-p)}{n}} = \sqrt{\frac{0.1(1-0.1)}{200}}$

$$= \sqrt{0.000\,45} = \underline{0.0212 \; (2.12\%)}$$

(b) Assuming a Normal distribution for the proportion of defectives in a sample, the 95% confidence limits are given by

$\mu_{\hat{p}} \pm 1.96 \sigma_{\hat{p}} = 0.1 \pm 1.96(0.0212)$

$= 0.1 \pm 0.0416$

$= \underline{0.0584 \text{ to } 0.1416}$

Thus the 95% confidence limits for the proportion of defectives in a sample are 5.84% to 14.16%.

EXAMPLE 3.6.3 In a random sample of 300 employees, 55% were found to be in favour of strike action. Find the 95% confidence limits for the proportion of all employees in the company who are in favour of such action.

SOLUTION We are given that $\hat{p} = 0.55$ and $n = 300$.

$$\therefore \quad \mu_{\hat{p}} \approx \hat{p} = 0.55$$

$$\sigma_{\hat{p}} \approx \sqrt{\frac{\hat{p}(1-\hat{p})}{n}} = \sqrt{\frac{0.55(0.45)}{300}}$$

$$= \sqrt{0.000\,825} = 0.028\,72$$

Assuming a Normal distribution the 95% confidence limits are

$$\mu_{\hat{p}} \pm 1.96\sigma_{\hat{p}} = 0.55 \pm 1.96(0.028\,72)$$

$$= 0.55 \pm 0.0563$$

$$= \underline{0.4937 \text{ to } 0.6063}$$

The 95% confidence limits are 49.37% to 60.63%.

EXAMPLE 3.6.4 25% of employees in a company are female. Find the probability that in a random sample of 120 employees less than 15% will be female.

SOLUTION We have $p = 0.25$ and $n = 120$.

$$\therefore \quad \mu_{\hat{p}} = p = 0.25$$

$$\sigma_p = \sqrt{\frac{p(1-p)}{n}} = \sqrt{\frac{0.25(0.75)}{120}} = 0.039\,53$$

Assuming a Normal distribution the required probability is shown below.

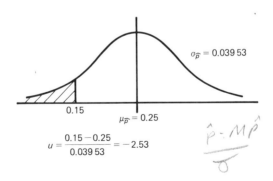

$$u = \frac{0.15 - 0.25}{0.039\,53} = -2.53$$

Required probability = area shaded = $\underline{0.0057}$ from tables.

EXAMPLE 3.6.5 It is assumed that over half of the employees in a large organisation are in favour of a proposed new wage structure. A random sample of 340 employees found that 56% were in favour. Does this sample verify the assumption? (Use the 5% significance level.)

SOLUTION We are given $\hat{p} = 0.56$, $n = 340$. $H_0: p = 0.50$, $H_1: p > 0.50$. Now

$$\sigma_{\hat{p}} = \sqrt{\frac{p(1-p)}{n}} = \sqrt{\frac{0.5(1-0.5)}{340}}$$

$$= \sqrt{0.0007352} = 0.02711$$

$$\therefore \quad u = \frac{\hat{p} - \mu_{\hat{p}}}{\sigma_{\hat{p}}} = \frac{0.56 - 0.50}{0.02711} = \underline{2.213}$$

A significant value of u at the 5% level (one tail) is 1.64. Therefore, our value of $u(= 2.213)$ is significant. We can reject H_0. We thus accept the assumption that the population does have a proportion greater than 50%, i.e. over half of the employees are in favour of the proposed new wage structure.

EXAMPLE 3.6.6 The following results have been recorded from random samples of candidates taking two Institute examinations.

Examination	No. of candidates sampled	No. of passes
Economics	48	22
Accountancy	82	31

Use the 5% level of significance to examine whether there is a significant difference in the proportions of candidates passing the two examinations.

SOLUTION We are given $n_1 = 48$, $\hat{p}_1 = 22/48 = 0.4583$ and $n_2 = 82$, $\hat{p}_2 = 31/82 = 0.3780$. $H_0: p_1 = p_2$, $H_1: p_1 \neq p_2$. Now

$$u = \frac{|\hat{p}_1 - \hat{p}_2|}{\sqrt{\hat{p}(1-\hat{p})(1/n_1 + 1/n_2)}}$$

where

$$\hat{p} = \frac{n_1\hat{p}_1 + n_2\hat{p}_2}{n_1 + n_2} = \frac{48(0.4583) + 82(0.3780)}{48 + 82}$$

$$= \frac{53}{130} = \underline{0.4077}$$

$$\therefore \quad u = \frac{0.4583 - 0.3780}{\sqrt{(0.4077)(1 - 0.4077)(1/48 + 1/82)}}$$

$$= \frac{0.0803}{\sqrt{0.007976}} = \underline{0.899}$$

A significant value of u at the 5% level (two tails) is 1.96. Therefore, our value of $u(= 0.899)$ is not significant. We accept H_0. There is thus no significant difference between the proportions of candidates passing the two examinations.

EXERCISES ON SECTION 3.6

1. Find the mean and standard deviation of the sampling distribution of proportions given that:

(a) $n = 100$ and $p = 0.5$,

(b) $n = 60$ and $p = 0.3$.

2. Given that $n = 250$ and $p = 0.4$ find the 95% confidence limits for the sample proportions.

3. During a given year 18% of employees were absent from work for more than three days due to illness. In a group of 140 employees find the probability that the proportion of employees absent for more than three days is (a) over 25%, and (b) less than 15%.

4. Out of a random sample of 320 borrowers from a large building society, 184 have mortgages in excess of £12 000. Find the 95% confidence limits for the proportion of such borrowers in the building society.

5. From a random sample of 85 UK companies it was found that 32 companies had annual turnovers in excess of one million pounds. Using a 5% significance level, test the assumption that 30% of all UK companies have over one million pounds annual turnover.

6. In a survey of customers' opinions, random samples of 50 customers were questioned at two stores. The results of the survey showed that 66% of customers were satisfied with the service in Store A compared with 54% in Store B. Use a 1% level of significance to test whether there is a significant difference between the proportion of satisfied customers in the two stores.

7. In a survey of voting intentions prior to a General Election, 35% from a random sample of 300 voters said that they intended to vote for the Conservative candidate. In a second area in the same constituency there were 42% intending to vote Conservative in a sample of 240. Use a 5% level of significance to determine whether there is a significant difference between the voting intentions in the two areas.

8. A quality inspection of two production lines gave the following results.

Production line	Sampling size	No. of defectives
A	40	3
B	25	6

Use a 5% significance level to examine the claim that line A is more reliable than line B.

PAST QUESTIONS 3

1. (a) The manufacturer of a certain oil-additive claims that the mean net weight of jars of his product is 1 kg. A random sample of size 49 of a large consignment supplied to your company is found to have a mean weight of 0.98 kg with a standard deviation of 0.02 kg. Test the manufacturer's claim at the 0.05 significance level.

(b) In an industrial process, two different methods are under review, both relating to the same job. 36 jobs are timed using Method 1 and 50 using Method 2, and the following results are obtained:

Method 1	Average job time = 54 minutes
	Standard deviation = 6 minutes
Method 2	Average job time = 57 minutes
	Standard deviation = 8 minutes

Do the methods yield significantly different average job times? [IIM]

2. (a) The data sheet published by a company states that the mean life of its X1 condenser is 1000 hours. A customer experiences a mean life of 970 hours and a standard deviation of 80 hours for a sample of 64 condensers. Determine if the customer's results show that the claimed mean life is false.

(b) An alarm system depends on the reaction of detectors and the probability that a given detector will react properly is known to be 0.9. In order to increase the reliability of the system, three detectors are mounted together, the alarm being set off if at least one detector reacts. Find the probability that the alarm will operate when required. [IIM]

3. (a) A food processing firm buys fresh fruit from an importer. The importer has claimed that as an overall average no more than 5% of any fruit delivery will be unfit for processing. The firm has inspected its most recent consignment of apples and discovered 40 bad ones in a total supply of 500. Should the firm reject the importer's claim? (You may use the normal approximation to the Binomial distribution.)

(b) An engineering company buys in a special component which in the past has had an overall mean length of 18.5 cm. The company's most recent delivery of 80 components had a mean length of 18.4 cm and standard deviation of 0.1 cm. Is this evidence of a real change in the component's mean length? [IIM]

4. There are two cranes, A and B, which attend to jobs as required. Using activity sampling to estimate their utilisation it was found that, out of 100 observations taken at random on each crane, A was working on 63 occasions and B on 47. It was expected that A and B would be equally utilised so these figures were considered somewhat surprising.

Required:

(a) What would you estimate as the percentage utilisation of each crane, assuming that A and B are equally utilised in the long run?

(b) Test whether the observations from activity sampling are consistent with A and B being equally utilised in the long run.

(c) What minimum difference would need to have been found between the observed percentage utilisations of A and B in order for a significant difference between their utilisations to be claimed at the 0.01 level?

[ACA]

5. (a) Discuss critically the following statement, made by a non-statistician, as fully as you can:

'In sampling theory the error involved when estimating a population parameter is inversely related to the square root of the sample size. The gain in accuracy obtained by increasing the sample fourfold is only twofold and it is therefore not worth the additional expense.'

(b) An assembly operation in a manufacturing plant requires a one month training period before a new operator can work it effectively. After a period of training an operator should be able to assemble a complete unit in 36 minutes, on average. This is the time allowed for in production calculations. A group of 9 new operators recorded the following times after they had completed the training period:

35, 30, 29, 28, 34, 36, 27, 32, 37 (all in minutes)

You are required to:

(i) Carry out a test of significance to find whether this group of operators indicates that the time of 36 minutes could be revised downwards for all new operators.

(ii) State any assumptions you have made in (i) and fully interpret your conclusion. [CIPFA]

6. (a) An automatic filling machine fills detergent bottles and it is known that 10% of bottles are underfilled. In a random sample of five bottles, what is the probability that:

(i) no bottles are underfilled?

(ii) at least two bottles are underfilled?

If an actual sample gave four bottles being underfilled, what conclusion would you draw?

(b) In a factory, components are received from two suppliers A and B. Using a quality control inspection scheme, random samples of 100 components were taken and showed that 10% of the components from supplier A were defective and that 6% of those from supplier B were defective. Apply an appropriate statistical test of significance to see if there is strong evidence that supplier B's quality is better than that of supplier A. [IIM]

7. (a) Explain fully the meaning of the standard error of a proportion, p, given by the formula

$$\sqrt{\frac{p(1-p)}{n}}$$

(b) (i) Calculate the probability that, if 30% of items in a batch are defective, a random sample of 100 items will contain 40% or more defectives.

(ii) It is claimed that a process produces not more than 30% defective. A random sample of 100 items from the process was found to contain 42 defectives. Investigate the validity of the claim. [ACA]

8. (a) Explain the basis of a test you would employ to determine the statistically significant difference between two sample proportions.

(b) In a survey of political opinions two samples, 3600 and 3000, were taken in 1965 and 1980. In 1965, 2000 respondents were in favour of the union with the European Economic Community while in 1980, 1600 respondents were in favour of the union. Is the difference in the proportions favouring the union in 1965 and 1980 statistically significant? Comment upon your results. [RVA]

9. Recruitment to a particular craft job in your firm is controlled by the relevant trade union. It is suggested that the policy of the union discriminates against non-white labour. Some 20% of applicants are non-white and on average there is no appreciable difference in qualifications or experience between white and non-white applicants. Of the last 225 persons recruited, 34 were non-white. Using the statistical theory of hypothesis testing determine whether there is any evidence of discrimination

   (a) if the possibility of bias in *preference* of non-whites is ruled out, and
   (b) if the possibility of either a positive or a negative bias is admitted. [IPM]

10. (a) Explain what is meant by (i) null hypothesis; (ii) significance level, in the context of a significance test.

    (b) A manufacturer claims that the mean breaking strength of a wire is at least 45 lb. A random sample of 100 pieces of the wire is tested and gives the following results:

Breaking strength (lb)	No. of pieces
30 but under 35	8
35 but under 40	21
40 but under 45	40
45 but under 50	23
50 but under 55	8

Do you consider the manufacturer's claim to be reasonable? [IIM]

11. Age of people consulting lawyers

Age in years	No. of people
18–24	190
25–29	263
30–34	286
35–44	417
45–54	350
55–64	244
65 and over	239

Estimate at a 99% level of confidence the average age of people consulting lawyers on the basis of the above random sample data. Discuss the basis of your estimation and express an opinion as to the value of your result. [RVA]

12. The standard time for a task includes a contingency allowance of 10 minutes per unit produced. A random sample of 50 units gives a distribution of actual time taken by contingencies as shown below. Show that the sample indicates a significant difference from the standard allowance.

Actual time for contingencies per unit (min)	Frequency
Less than 1	2
1– 3	8
3– 5	10
5–10	14
10–15	9
15–20	6
20–25	1

[IMS]

13. Time studies of two groups of workers in different locations produced the following results:

Group	Standard time	No. of observations	Variance
L	32.62 min	15	1.89
M	34.25 min	9	3.42

Test the data statistically and comment on the results. [IMS]

14. A life office has introduced a facility providing on-line access to its main computer record file to one branch for a trial over one quarter. The facility allows surrender value and paid-up policy quotations to be calculated on the spot rather than through written requests to head office as under the previous system.

The office has however become concerned that the proportion of surrendered and paid-up policies arising from this branch has increased over the period.

The quarter's figures for the branch are as follows:

Average number of policies in force	10 000
No. of surrender values taken	1000
No. of policies made paid-up	400

The office as a whole is accustomed to a surrender rate of 9.5% of the average number of policies in force each year and a paid-up policy rate of 3.8% per year. These rates have been based on a very large number of policies in force compared with the branch in question.

(a) Do you consider that the experience of this branch is significantly different from that of the office as a whole over the quarter? You should treat surrendered and paid-up policies separately.

(b) Would your view change if identical figures were shown over the next quarter?

(c) What additional information (if any) would you consider useful in order to determine whether or not the introduction of the facility had adversely affected the branch's experience in this respect? [CII]

4 THE χ^2 DISTRIBUTION

4.1 GOODNESS OF FIT

$$\chi^2 = \sum \frac{(O-E)^2}{E}$$

where $O =$ observed frequency, $E =$ expected frequency. The number of degrees of freedom $\nu = n - r$ ($n =$ number of classes, $r =$ number of restrictions).

EXAMPLE 4.1.1 From tables, find the 95% confidence limits for χ^2 with $\nu = 5$.

SOLUTION

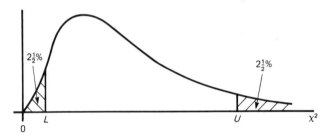

The 95% limits, L and U, obtained from tables are 0.831 212 to 12.8325.

EXAMPLE 4.1.2 Find a 'significant' value of χ^2 at the 5% level with the number of degrees of freedom given by (a) $\nu = 4$, and (b) $\nu = 10$.

SOLUTION

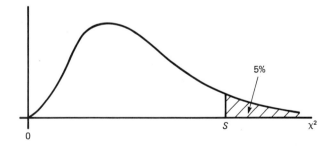

S is a significant value at the 5% level.
(a) When $\nu = 4$, the 5% significant value is 9.487 73.
(b) When $\nu = 10$, the 5% significant value is 18.3070.

EXAMPLE 4.1.3 Use the χ^2 test at the 1% level of significance with $\nu = 3$ to investigate whether there is any significant difference between the observed and expected frequencies given below:

Observed	12	19	26	43
Expected	10	20	30	40

SOLUTION The value of χ^2 is calculated in the table below.

O	E	O−E	(O−E)²	(O−E)²/E
12	10	2	4	0.4
19	20	−1	1	0.05
26	30	−4	16	0.533
43	40	3	9	0.225
				$\chi^2 = 1.208$

Now, a significant value of χ^2 at the 1% level with $\nu = 3$ is 11.3449. Therefore, our value of $\chi^2 (= 1.208)$ is less than this and is not significant. We thus conclude that there is no significant difference between the observed and expected frequencies.

EXAMPLE 4.1.4 In an experiment a die is tossed 60 times yielding the following results.

Score	1	2	3	4	5	6
Frequency	9	6	8	11	7	19

Use the χ^2 test at the 5% level of significance to investigate whether the die is biased.

SOLUTION Assuming the die is unbiased the expected frequency distribution is given below:

Score	1	2	3	4	5	6
Frequency	10	10	10	10	10	10

Comparing the observed and expected frequencies we have:

O	E	O−E	(O−E)²	(O−E)²/E
9	10	−1	1	0.1
6	10	−4	16	1.6
8	10	−2	4	0.4
11	10	1	1	0.1
7	10	−3	9	0.9
19	10	9	81	8.1
				$\chi^2 = 11.2$

Now, in this example, $n = 6$ and $r = 1$.

[*Note*: The one and only restriction on the calculation of E is the *total*, i.e. the total of the expected frequencies must be equal to 60.]

$$\therefore \quad \nu = n - r = 6 - 1 = 5$$

From tables, a significant value of χ^2 at the 5% level with $\nu = 5$ is 11.0705. Therefore, our value of $\chi^2 (= 11.2)$ is greater than this and is significant. We reject the assumption. The evidence thus seems to indicate that the die is biased.

EXAMPLE 4.1.5 The number of deliveries per day arriving at a warehouse over a period of 50 working days is given in the table below.

No. of deliveries	0	1	2	3	4 or more
No. of days	11	18	10	7	4

Use the χ^2 test at the 5% level to investigate whether the number of deliveries conforms to the Poisson distribution.

SOLUTION Assuming that the number of deliveries does conform to the Poisson distribution we have:

$$p(r) = \frac{e^{-\lambda} \cdot \lambda^r}{r!}$$

where λ = mean. Now, we can obtain λ from the observed distribution:

x	0	1	2	3	4*	
f	11	18	10	7	4	$\Sigma f = 50$
fx	0	18	20	21	16	$\Sigma fx = 75$

[*Note: Strictly, we should use the midpoint of the class '4 or more'. However, using $x = 4$ in this case will not seriously affect the required calculations.]

$$\therefore \quad \text{Mean} = \frac{\Sigma fx}{\Sigma f} = \frac{75}{50} = 1.5, \quad \text{i.e.} \quad \lambda = 1.5$$

$$\therefore \quad p(0) = \frac{e^{-1.5}(1.5)^0}{0!} = 0.223\,13$$

$$p(1) = \frac{e^{-1.5}(1.5)^1}{1!} = 0.334\,70$$

$$p(2) = \frac{e^{-1.5}(1.5)^2}{2!} = 0.251\,02$$

$$p(3) = \frac{e^{-1.5}(1.5)^3}{3!} = 0.125\,51$$

$$p(4 \text{ or more}) = 1 - [p(0) + p(1) + p(2) + p(3)]$$
$$= 1 - 0.934\,36 = 0.065\,64$$

The theoretical Poisson probability distribution is thus:

x	0	1	2	3	4 or more
$p(x)$	0.223 13	0.334 70	0.251 02	0.125 51	0.065 64

THE χ^2 DISTRIBUTION

The expected frequency distribution is:

x	0	1	2	3	4 or more
f	11.16	16.73	12.55	6.28	3.28

Comparing the observed and expected frequencies we have:

O	E	$O-E$	$(O-E)^2$	$(O-E)^2/E$
11	11.16	-0.16	0.0256	0.0023
18	16.73	1.27	1.6129	0.0964
10	12.55	-2.55	6.5025	0.5181
11 $\begin{cases} 7 \\ 4 \end{cases}$	$\begin{cases} 6.28 \\ 3.28 \end{cases}$ 9.56*	1.44	2.0736	0.2169
				$\chi^2 = 0.8337$

[*Note: The χ^2 test is unreliable for small *expected* frequencies (i.e. E less than 5). Consequently the last two classes are combined.]

In this example we have $n = 4$ and $r = 2$. (The *total* and *mean* are the two restrictions on the expected frequencies.)

From tables, a significant value of χ^2 at the 5% level with $v = n - r = 4 - 2 = 2$ is 5.99. Therefore, our value of $\chi^2 (= 0.8337)$ is not significant. From the observed data we thus conclude that the number of deliveries per day is likely to conform to the Poisson distribution.

EXAMPLE 4.1.6 The table below gives the wages of a sample of employees in a large organisation.

Weekly wage (£)	80–	100–	120–	140–	160–	180–
No. of employees	10	25	45	10	5	5

Test the assumption that wages in this organisation conform to the Normal distribution.

SOLUTION Assuming that the wages are Normally distributed we require the mean (μ) and standard deviation (σ) of the observed distribution.

We are given:

x(midpoint)	90	110	130	150	170	190
f	10	25	45	10	5	5

We find that $\Sigma fx = 12\,800$ and $\Sigma fx^2 = 1\,694\,000$.

$$\therefore \quad \mu = \frac{\Sigma fx}{\Sigma f} = \frac{12\,800}{100} = 128$$

$$\sigma = \sqrt{\frac{\Sigma fx^2}{\Sigma f} - \left(\frac{\Sigma fx}{\Sigma f}\right)^2} = \sqrt{556} = 23.58$$

To evaluate the expected frequencies we find areas under the Normal curve with $\mu = 128$ and $\sigma = 23.58$ between the limits given in the observed distribution.

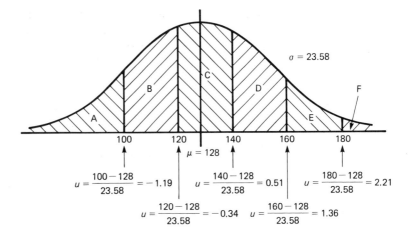

$$u = \frac{100-128}{23.58} = -1.19 \quad u = \frac{140-128}{23.58} = 0.51 \quad u = \frac{180-128}{23.58} = 2.21$$

$$u = \frac{120-128}{23.58} = -0.34 \quad u = \frac{160-128}{23.58} = 1.36$$

Using Normal tables, we find that area A = 0.117 02, area B = 0.249 91, area C = 0.328 04, area D = 0.218 11, area E = 0.073 37, area F = 0.013 55.

To obtain the expected frequencies we multiply these probabilities by 100.

x	below 100	100–	120–	140–	160–	180–
f	11.702	24.991	32.804	21.811	7.337	1.355

Comparing the observed and expected frequencies we have:

O	E	O−E	(O−E)²	(O−E)²/E
10	11.702	−1.702	2.8969	0.2475
25	24.991	0.009	0.0001	0.0000
45	32.804	12.196	148.7424	4.5343
10	21.811	−11.811	139.4997	6.3958
10 { 5 / 5 }	{ 7.337 / 1.355 } 8.692	1.308	1.7109	0.1968
				$\chi^2 = 11.3745$

We have $n = 5$ and $r = 3$ (the *total*, *mean* and *standard deviation* are restrictions on the Normal expected frequencies).

$$\therefore \quad \nu = n - r = 5 - 3 = 2$$

Now, a significant value of χ^2 at the 5% level with $\nu = 2$ is 5.99. Therefore, our value of $\chi^2 (= 11.3745)$ is significant. We thus reject the assumption, i.e. it is unlikely that wages in the organisation are Normally distributed.

EXERCISES ON SECTION 4.1

1. Find a significant value of χ^2 at the 5% level with:
   (a) $\nu = 6$,
   (b) $\nu = 15$,
   (c) $\nu = 30$.

2. Find the 99% confidence limits for χ^2 with the number of degrees of freedom = 4.

3. Assuming that $\nu = n-1$ use the χ^2 test at the 5% level of significance to investigate whether there is a significant difference between the following sets of observed and expected frequencies.

(a)	O	8	10	12	10	8	
	E	11	9	10	9	11	

(b)	O	20	30	15	10	5
	E	15	28	18	10	9

(c)	O	4	12	19	15	10	10
	E	6	18	14	19	9	4

4. A random sample of 200 employees in a large company has the age distribution given below.

Age (years)	Under 25	25–	40–	55–
No. of employees	42	82	31	45

Use the χ^2 test at the 5% level to examine the assumption that the distribution of ages in the company is as shown below.

Age (years)	Under 25	25–	40–	55–
Percentage	15	40	25	20

5. The number of industrial accidents per month occurring in a large manufacturing company over the past 10 years is given in the table below.

No. of accidents	0	1	2	3	4	5
No. of months	20	27	30	25	12	6

Use the χ^2 test at the 5% level to examine whether the number of accidents conforms to the Poisson distribution.

6. Use the χ^2 test at the 1% level of significance to examine whether survival times in an organisation are Normally distributed based on the following sample.

Length of service (years)	0–	5–	10–	15–	20–	25–
No. of employees	8	24	30	19	6	3

4.2 CONTINGENCY TABLES

In an $m \times n$ contingency table the number of degrees of freedom
$$\nu = (m-1)(n-1)$$

EXAMPLE 4.2.1 The contingency table below shows the observed 'X' and 'Y' values obtained in a sample of 100 items.

Values of Y	Values of X	
	High	Low
High	30	10
Low	25	35

Assuming that there is no association between the values of X and Y find the expected frequencies in this table.

SOLUTION We are given the observed frequencies as shown below.

30	10	40
25	35	60
55	45	100

Assuming no association between X and Y the expected frequencies are calculated in the following table.

$\dfrac{55 \times 40}{100} = 22$	$\dfrac{45 \times 40}{100} = 18$	40
$\dfrac{55 \times 60}{100} = 33$	$\dfrac{45 \times 60}{100} = 27$	60
55	45	100

[*Note*: Each expected frequency is calculated by the formula
(column total) × (row total) ÷ (grand total).]

EXAMPLE 4.2.2 State the number of degrees of freedom in the table given in Example 4.2.1.

SOLUTION We are given a 2 × 2 table (i.e. 2 rows and 2 columns).

∴ $\nu = (2-1)(2-1) = (1)(1) = 1$

∴ No. of degrees of freedom = 1

EXAMPLE 4.2.3 The following table shows the relationship between the assessments of job performance (based on an annual appraisal) and educational background of a sample of 80 employees in middle management in a large organisation.

Assessment	Educational status	
	Graduates	Non-graduates
Good	19	9
Average	20	16
Poor	11	5

Use the χ^2 test at the 5% level to test whether there is any association between the educational background and assessments of job performance of middle managers in this organisation.

SOLUTION The observed frequencies are shown in the table below.

19	9	28
20	16	36
11	5	16
50	30	80

Assuming that there is no association between performance appraisals and educational status, the expected frequencies are calculated below.

$\dfrac{50 \times 28}{80} = 17.5$	$\dfrac{30 \times 28}{80} = 10.5$	28
$\dfrac{50 \times 36}{80} = 22.5$	$\dfrac{30 \times 36}{80} = 13.5$	36
$\dfrac{50 \times 16}{80} = 10$	$\dfrac{30 \times 16}{80} = 6$	16
50	30	80

Comparing the observed and expected frequencies we have:

O	E	O−E	(O−E)²	(O−E)²/E
19	17.5	1.5	2.25	0.1286
9	10.5	−1.5	2.25	0.2143
20	22.5	−2.5	6.25	0.2778
16	13.5	2.5	6.25	0.4630
11	10	1	1	0.1
5	6	−1	1	0.1667
				$\chi^2 = 1.3504$

We have a 3 × 2 contingency table

$$\therefore \quad v = (3-1)(2-1) = (2)(1) = 2$$

Now a significant value of χ^2 at the 5% level with $v = 2$ is given by 5.991 46. Therefore, our value of $\chi^2 (= 1.3504)$ is less than this and is not significant. We accept the assumption. It is thus unlikely that there is any association between the job performance assessments and educational backgrounds of the employees.

EXAMPLE 4.2.4 A wholesaler receives mass-produced parts from three different manufacturers. The table below shows the quality of batches obtained from each manufacturer over the past three months.

Percentage of imperfect items	Manufacturer			Total
	A	B	C	
Under 1%	15	30	25	70
1–2%	10	20	40	70
2–4%	12	10	18	40
Over 4%	3	10	7	20
Total	40	70	90	200

Use the χ^2 test at the 1% level to find whether there is any significant difference between the quality of goods obtained from the three manufacturers.

SOLUTION We are given the observed frequencies.

15	30	25	70
10	20	40	70
12	10	18	40
3	10	7	20
40	70	90	200

Assuming that there is no relationship between the manufacturers and the quality of goods received, the expected frequencies are calculated below.

$\dfrac{40 \times 70}{200} = 14$	$\dfrac{70 \times 70}{200} = 24.5$	$\dfrac{90 \times 70}{200} = 31.5$	70
$\dfrac{40 \times 70}{200} = 14$	$\dfrac{70 \times 70}{200} = 24.5$	$\dfrac{90 \times 70}{200} = 31.5$	70
$\left.\begin{array}{l}\dfrac{40 \times 40}{200} = 8 \\ \dfrac{40 \times 20}{200} = 4\end{array}\right\}12$	$\left.\begin{array}{l}\dfrac{70 \times 40}{200} = 14 \\ \dfrac{70 \times 20}{200} = 7\end{array}\right\}21$	$\left.\begin{array}{l}\dfrac{90 \times 40}{200} = 18 \\ \dfrac{90 \times 20}{200} = 9\end{array}\right\}27$	40 20
40	70	90	200

[*Note*: We combine the third and fourth rows together to avoid small values of E (i.e. less than 5) in the χ^2 test.]

Comparing the observed and expected frequencies we have:

	O	E		$O-E$	$(O-E)^2$	$(O-E)^2/E$
	15	14		1	1	0.0714
	10	14		-4	16	1.1429
15	$\left\{\begin{array}{c}12\\3\end{array}\right.$	$\left.\begin{array}{c}8\\4\end{array}\right\}$	12	3	9	0.75
	30	24.5		5.5	30.25	1.2347
	20	24.5		-4.5	20.25	0.8265
20	$\left\{\begin{array}{c}10\\10\end{array}\right.$	$\left.\begin{array}{c}14\\7\end{array}\right\}$	21	-1	1	0.0476
	25	31.5		-6.5	42.25	1.3413
	40	31.5		8.5	72.25	2.2937
25	$\left\{\begin{array}{c}18\\7\end{array}\right.$	$\left.\begin{array}{c}18\\9\end{array}\right\}$	27	-2	4	0.1481
						$\chi^2 = 7.8562$

We have a 3×3 contingency table.

[*Note*: The third and fourth rows are combined.]

$$\therefore \quad \nu = (3-1)(3-1) = (2)(2) = 4$$

Now a significant value of χ^2 at the 1% level with $\nu = 4$ is 13.2767. Therefore, our value of $\chi^2 (= 7.8562)$ is not significant. We accept the assumption.

From the data obtained over the past three months there thus seems to be no significant difference between the quality of goods supplied by the three manufacturers.

4.3 YATES' CORRECTION

When $\nu = 1$ it is necessary to use Yates' correction

$$\chi^2 = \sum \frac{Y^2}{E}$$

where $Y = |O - E| - \frac{1}{2}$.

EXAMPLE 4.3.1 Use the χ^2 test at the 1% level to examine whether there is any association between the values of A and B given the sample tabulated below.

Values of B	Values of A	
	High	Low
High	14	16
Low	6	64

SOLUTION The observed frequencies are:

14	16	30	
6	64	70	
20	80	100	

Assuming that there is no association between A and B the expected frequencies are as follows.

$\frac{20 \times 30}{100} = 6$	$\frac{80 \times 30}{100} = 24$	30
$\frac{20 \times 70}{100} = 14$	$\frac{80 \times 70}{100} = 56$	70
20	80	100

Now, in a 2×2 contingency table $\nu = (2-1)(2-1) = (1)(1) = 1$. It is therefore necessary to use Yates' correction in the calculation of χ^2 given below.

O	E	$O - E$	$Y = |O - E| - \frac{1}{2}$	Y^2	Y^2/E
14	6	8	7.5	56.25	9.375
16	24	−8	7.5	56.25	2.344
6	14	−8	7.5	56.25	4.018
64	56	8	7.5	56.25	1.004
					$\chi^2 = 16.741$

A significant value of χ^2 at the 1% level with $\nu = 1$ is 6.6349. Therefore, our value of $\chi^2 (= 16.741)$ is significant. We reject the assumption. The given sample thus indicates some association between A and B.

EXAMPLE 4.3.2 The following table shows the travelling distance to work together with the general opinion of working conditions of a sample of 160 workers.

Distance travelled to work	Attitude towards working conditions	
	Satisfied	Dissatisfied
Over 10 miles	68	22
Under 10 miles	62	8

Investigate the relationship between travelling distances to work and employees' attitudes based on this sample.

SOLUTION The observed frequencies are:

68	22		90
62	8		70
130	30		160

Assuming that there is no relationship between the attitudes and travelling distances we have the expected frequencies given in the table below:

73.125	16.875		90
56.875	13.125		70
130	30		160

[*Note*: $\nu = 1$.]

Comparing the observed and expected frequencies we have:

O	E	$O-E$	$Y = \lvert O-E \rvert - \frac{1}{2}$	Y^2	Y^2/E
68	73.125	-5.125	4.625	21.3906	0.2925
22	16.875	5.125	4.625	21.3906	1.2676
62	56.875	5.125	4.625	21.3906	0.3761
8	13.125	-5.125	4.625	21.3906	1.6298
					$\chi^2 = 3.5660$

Now a significant value of χ^2 at the 5% level with $\nu = 1$ is 3.841 46.

[*Note*: When the significance level is not specified in a question it is entirely a matter of choice. However, the 5% level is used more often than not — at least, in examination questions, if not in real life!]

Therefore, our value of $\chi^2 (= 3.5660)$ is not significant. We accept the assumption. There is thus no evidence of an association between employee attitudes towards working conditions and travelling distances to work.

EXERCISES ON SECTIONS 4.2 AND 4.3

1. State the number of degrees of freedom involved in evaluating expected frequencies given the observed frequencies in a contingency table of size (*a*) 3 × 2, (*b*) 5 × 3 and (*c*) 5 × 6.

2. Use a χ^2 test at the 5% level of significance to investigate whether there is any association between the values of X and Y given in each of the contingency tables given below.

(a)

Y values	X values		
	High	Medium	Low
High	15	13	12
Low	15	17	28

(b)

Y values	X values			
	Below 100	100–120	120–150	Above 150
Below 60	37	17	8	8
60–100	54	28	12	6
Above 100	9	5	10	6

(c)

Y values	X values	
	Below 100	100 or more
Below 100	28	74
100 or more	36	42

3. A nationwide TV rental company has conducted a survey into customer approval of a number of different makes of television. The table below gives the responses of a random sample of 400 customers from various retail outlets. Use the χ^2 test at the 5% level to investigate whether there is any significant difference between the different makes of television in the opinions of the customers.

Opinions of customers	Type of TV				Total
	A	B	C	D	
Satisfied	83	60	57	50	250
Dissatisfied	22	26	16	26	90
Total	105	86	73	76	340

4. The table below shows the absentee records of a sample of unskilled employees from three companies.

Company	Absentee records		
	Good	Average	Poor
A	27	37	6
B	20	24	6
C	4	14	12

Use the χ^2 test at the 1% level of significance to investigate whether there is any significant difference in the absenteeism in the three companies.

5. The selection procedure for new recruits in a company consists of two stages. The potential recruits are given an aptitude test followed by an interview by a panel. The table shows the results of each stage of the selection procedure for a sample of applicants taken over the past five years. Is there evidence of an association between the recommendations from the two stages?

	Selection stage	
Assessment	Test	Interview
Satisfactory	46	30
Unsatisfactory	38	54

PAST QUESTIONS 4

1. Three new types of diesel-electric locomotives were tested under the same conditions for periods of 10, 8 and 6 weeks respectively, and the number of breakdowns which occurred were 20, 7 and 9 respectively.

   Do these figures indicate a significant difference in quality between the locomotives?

   Explain the practical applications of the chi-squared test, giving a few examples. What are its restrictions and its advantages? [CIT]

2. A large tutorial school has three lecturers, Smith, Jones and Brown, who prepare different groups of students for the same examination, each group being of the same average ability. Examination results are graded as either distinction, pass or fail, and the results obtained by 100 students are classified below according to lecturer and grade obtained.

	Smith	Jones	Brown
Distinction	5	4	8
Pass	13	16	14
Fail	12	20	8

   Required:

   (a) Use a suitable test of significance to compare the results obtained by the three lecturers and comment on your findings.

   (b) Suppose that, of a large number of students not attending the tutorial school, 11% obtained distinction, 31% passed and 58% failed. Assess whether there is any evidence that a student's performance is improved by attending the tutorial school and state a reservation you might have about your conclusion. [ACA]

3. In order to find out whether or not the spare wheel is exchanged regularly with the other wheels on the cars of salesmen, 1950 tyres were inspected, with the following results.

	Nos. in categories			
	New	Part-worn	Worn	Badly worn
Road wheels	338	874	278	70
Spare wheel	151	111	83	45

(a) Write a short report on your analysis and conclusions.

(b) Explain the advantages of the chi-squared test and name the two main categories of its applications. Are there any restrictions? [CIT]

4. A garage sells three types of new car, the Tuxan, Firedash and Velotta. The following data show, for each type of car sold, the number requiring repair during the first 12 months.

	No. of cars sold		Total
	Requiring repair	Not requiring repair	
Tuxan	89	11	100
Firedash	57	13	70
Velotta	118	12	130

Required:

(a) State the main feature indicated by these data.

(b) Set up an appropriate null hypothesis and test it.

(c) Explain your conclusions to the garage management. [ACA]

5. A study of a random sample of 500 undergraduates in higher education showed that they had come from the following social classes.

Parental occupation	Type of institution		
	Oxbridge	Other university	Polytechnic
Professional	105	70	85
Managerial	44	31	45
Manual	36	29	55

(a) Examine the proposition that the differences in the class composition of undergraduates at these various types of institution are statistically significant.

(b) Interpret your results and explain the reasoning behind the test you chose. [RVA]

6. In a survey of people in a particular locality an attempt was made to find out whether the frequency of visits to a doctor's surgery was associated with the age of respondents. Of 186 people interviewed the following results were obtained.

Age of respondent (in years)	Frequency of visiting doctor				Total
	Once a month	Every 6 months	Once a year	Less than once a year	
20 or less	5	6	10	16	37
21–40	12	18	8	33	71
41–60	20	11	2	10	43
Over 60	22	8	1	4	35
Total	59	43	21	63	186

(a) Carry out an appropriate statistical test to assess the significance of these results.

(b) Explain fully the theory underlying the test you have carried out and what conclusions you can come to as a result of the test. [CIPFA]

7. The performance of samples of managers who attended three different management training programmes, was followed up after a three-month interval, in order to ascertain the usefulness of the programmes. The table shows the performance rating rates (1 — excellent, 2 — good, 3 — moderate, 4 — unsatisfactory).

Programme	No. of managers in each performance rating grade			
	1	2	3	4
1	10	23	13	5
2	50	58	28	15
3	15	20	8	7

From your analysis of the above data, what conclusions would you draw about the comparative effectiveness of the three programmes? [IPM]

8. (a) In order to establish the popularity of two new designs of uniform, 40 station staff were divided randomly into two groups of 20. After a one-month trial period, they described their experiences as follows.

	Design A	Design B
Satisfactory	15	8
Unsatisfactory	5	12

What conclusions can be drawn?

(b) (i) What is the difference between parametric and non-parametric statistics?

(ii) Mention a few non-parametric tests and give examples of their application in practice. [CIT]

9. How ratepayers obtained information about the operation of the rating system, by social class:

How information obtained	Social class				
	Professional	Other non-manual	Skilled manual	Partly skilled	Unskilled
Mainly asked	15	95	114	58	20
Mainly told	7	74	102	64	25

Use a chi-squared test to see if there is any association between social class and the way information is obtained about the rating system.

Discuss your result and the basis of the test you have applied. [RVA]

10. A company decides to reconsider its recruitment policy with a view to reducing absence from work. Employment records for the previous year were examined and the results were summarised as follows.

| | No. of days absent | | |
Travelling time	Low	Medium	High
Low	31	24	20
Medium	24	26	35
High	18	32	47

(a) What is meant by the following terms in relation to this information?
  (i) Null hypothesis
  (ii) Alternative hypothesis
  (iii) Test statistic
  (iv) Significance level
  (v) Critical region.

(b) What conclusions do you draw from this evidence about the association between absence and travelling time, and why? [IPM]

11. Analysis of authorities' performance against targets of Local Authority expenditure:

Class of authority	No. in sample	No. on target	No. over target	No. under target
London Boroughs	20	1	16	3
Metropolitan Districts	25	4	14	7
Non-Metropolitan Districts	223	17	52	154
Welsh Districts	26	1	14	11
Non-Metropolitan Counties	39	13	12	14
Total	333	36	108	189

(*Source*: Rating and Valuation – May 1982)

Use a chi-squared test at a 95% and a 99% level of confidence to show if there are any differences in the performance against targets by class of authority. Interpret your answer. [RVA]

12. A random sample survey of 200 party workers produced the following information about their religious attitudes.

Religious affiliation	Labour	Conservative	Liberal
Catholic	40	30	5
Protestant	65	50	10

Test whether there is any association between political party support and religious affiliation at the 95% level of confidence. Interpret your result. [RVA]

13. A consignment of HD aluminium alloy tube is thought to have a mass/metre that is uniformly distributed in the range 3 to 4 kg/m. 100 measurements gave the following distribution when classified.

Mass/metre (kg)	3–3.2	3.2–3.4	3.4–3.6	3.6–3.8	3.8–4.0
Observed frequency	16	23	25	19	17

Formulate a null hypothesis to calculate expected frequencies, in order to check whether there is any evidence here that these data follow a uniform distribution (use χ^2 statistic at 5% level).

14. The number of faults requiring attention in a sample of 100 heavy duty deck hatches were found on inspection to be as follows, with mean 0.77.

No. of faults	0	1	2	3 or more	
Frequency	48	32	17	3	(total 100)

Do these data constitute evidence that the faults have a Poisson distribution with the same mean? In order to test this, formulate a null hypothesis to calculate expected frequencies, and carry out a χ^2 goodness-of-fit test at the 5% level.

5 CORRELATION AND REGRESSION

5.1 SCATTER DIAGRAMS

EXAMPLE 5.1.1 Illustrate the relationship between the intelligence (as measured by an IQ test) and earnings for the sample of 12 employees in a small company shown below.

Employee	A	B	C	D	E	F	G	H	I	J	K	L
Weekly wage (£)	75	78	86	88	90	95	95	98	106	110	112	122
IQ	102	96	98	103	101	100	105	109	105	105	110	120

SOLUTION The data are illustrated in the following scatter diagram.

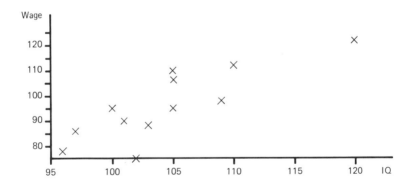

EXAMPLE 5.1.2 By drawing a suitable diagram illustrate the sales of an electronics company over the past 10 years.

Year	1973	1974	1975	1976	1977	1978	1979	1980	1981	1982
Sales (£1000)	23.5	25.4	24.7	27.8	31.2	36.4	42.1	39.8	44.4	50.6

SOLUTION

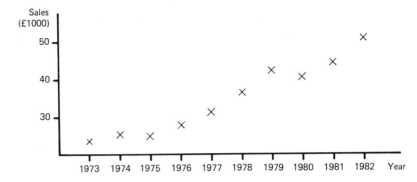

5.2 A CORRELATION COEFFICIENT

One method of investigating whether a linear relationship exists between two variables x and y is by calculating the *product moment correlation coefficient* (PMCC) denoted by r, and given by the formula:

$$r = \frac{\Sigma xy - n\bar{x}\bar{y}}{\sqrt{(\Sigma x^2 - n\bar{x}^2)(\Sigma y^2 - n\bar{y}^2)}}$$

where $-1 \leqslant r \leqslant 1$.

The following scatter diagrams illustrate certain values of the correlation coefficient.

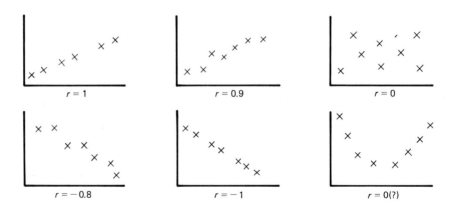

EXAMPLE 5.2.1 By calculating the PMCC find the degree of association between weekly earnings and the amount of income tax paid for each member of a group of 10 manual workers.

Weekly wage (£)	76	78	84	85	88	89	95	95	100	110
Income tax (£)	8	6	12	12	15	10	16	20	19	22

CORRELATION AND REGRESSION 93

SOLUTION The PMCC is calculated in the table below.

x	y	x^2	y^2	xy
76	8	5766	64	608
78	6	6084	36	463
84	12	7056	144	1008
85	12	7225	144	1020
88	15	7744	225	1320
89	10	7921	100	890
95	16	9025	256	1520
95	20	9025	400	1900
100	19	10 000	361	1900
110	22	12 100	484	2420
$\Sigma x = 900$	$\Sigma y = 140$	$\Sigma x^2 = 81\,956$	$\Sigma y^2 = 2214$	$\Sigma xy = 13\,054$

$$\bar{x} = \frac{\Sigma x}{n} = \frac{900}{10} = 90, \qquad \bar{y} = \frac{\Sigma y}{n} = \frac{140}{10} = 14$$

$$r = \frac{\Sigma xy - n\bar{x}\bar{y}}{\sqrt{[\Sigma x^2 - n\bar{x}^2][\Sigma y^2 - n\bar{y}^2]}}$$

$$= \frac{13\,054 - 10(90)(14)}{\sqrt{[81\,956 - 10(90)^2][2214 - 10(14)^2]}}$$

$$= \frac{13\,054 - 12\,600}{\sqrt{[81\,956 - 81\,000][2214 - 1960]}} = \frac{454}{\sqrt{(956)(254)}}$$

$$= \frac{454}{\sqrt{242\,824}} = \frac{454}{492.77} = \underline{0.921}$$

r is 'near' 1 and indicates a good positive linear correlation between the two variables.

EXAMPLE 5.2.2 Investigate the degree of correlation between the grades obtained in two assessment tests by a group of eight participants.

Test A	7	7	10	12	15	16	18	19
Test B	14	15	12	12	8	9	10	8

SOLUTION The calculation of r is tabulated below.

x	y	x^2	y^2	xy
7	14	49	196	98
7	15	49	225	105
10	12	100	144	120
12	12	144	144	144
15	8	225	64	120
16	9	256	81	144
18	10	324	100	180
19	8	361	64	152
$\Sigma x = 104$	$\Sigma y = 88$	$\Sigma x^2 = 1508$	$\Sigma y^2 = 1018$	$\Sigma xy = 1063$

94 NOTES AND PROBLEMS IN STATISTICS

$$\bar{x} = \frac{\Sigma x}{n} = \frac{104}{8} = 13, \qquad \bar{y} = \frac{\Sigma y}{n} = \frac{88}{8} = 11$$

$$r = \frac{\Sigma xy - n\bar{x}\bar{y}}{\sqrt{[\Sigma x^2 - n\bar{x}^2][\Sigma y^2 - n\bar{y}^2]}} = \frac{1063 - 8(13)(11)}{\sqrt{[1508 - 8(13)^2][1018 - 8(11)^2]}}$$

$$= \frac{1063 - 1144}{\sqrt{(156)(50)}} = \frac{-81}{\sqrt{7800}}$$

$$= \frac{-81}{88.3176}$$

$$r = \underline{-0.917}$$

r is 'near' -1 and indicates a good inverse linear correlation between the two variables.

EXAMPLE 5.2.3 Evaluate the PMCC for the following data.

x	10	15	20	25	30
y	148	146	149	154	153

SOLUTION [*Note*: Coding the values of x and/or y does *not* change the value of the PMCC obtained.]

A suitable coding and all necessary calculations are given below.

x	y	$X = \dfrac{x-10}{5}$	$Y = y - 146$	X^2	Y^2	XY
10	148	0	2	0	4	0
15	146	1	0	1	0	0
20	149	2	3	4	9	6
25	154	3	8	9	64	24
30	153	4	7	16	49	28
		10	20	30	126	58

$$\bar{X} = \frac{\Sigma X}{n} = \frac{10}{5} = 2, \qquad \bar{Y} = \frac{\Sigma Y}{n} = \frac{20}{5} = 4$$

$$\therefore \quad r = \frac{\Sigma XY - n\bar{X}\bar{Y}}{\sqrt{[\Sigma X^2 - n\bar{X}^2][\Sigma Y^2 - n\bar{Y}^2]}}$$

$$= \frac{58 - 5(2)(4)}{\sqrt{[30 - 5(2)^2][126 - 5(4)^2]}}$$

$$= \frac{18}{\sqrt{(10)(46)}} = \frac{18}{\sqrt{460}}$$

$$= \frac{18}{21.4476} = \underline{0.839}$$

5.3 RANK CORRELATION COEFFICIENT

The PMCC of ranks, known as the *rank correlation coefficient*, can be obtained by the formula,

$$r = 1 - \frac{6\Sigma d^2}{n(n^2-1)}$$

where d = difference between rankings.

EXAMPLE 5.3.1 Two members of an interview panel have ranked six applicants in order of preference for a specified post. Calculate the degree of agreement between the two members.

Applicant	A	B	C	D	E	F
Interviewer X	1	2	3	4	5	6
Interviewer Y	3	1	4	2	5	6

SOLUTION The differences in rankings are shown below.

d	2	1	1	2	0	0
d^2	4	1	1	4	0	0

$\Sigma d^2 = 10$

$r = 1 - \dfrac{6\Sigma d^2}{n(n^2-1)}$

$= 1 - \dfrac{6(10)}{6(36-1)}$

$= 1 - \dfrac{60}{210}$

$= 1 - 0.2857$

$r = \underline{0.7143}$

EXAMPLE 5.3.2 The results of two tests taken by eight employees are shown below (figures in %).

Employee	A	B	C	D	E	F	G	H
Test X	48	50	56	64	68	72	75	84
Test Y	54	49	51	63	65	62	79	74

Rank each employee in order of performance in the two tests and calculate the rank correlation coefficient.

SOLUTION Ranking the employees in each test we have:

Employee	A	B	C	D	E	F	G	H
Rank in X	8	7	6	5	4	3	2	1
Rank in Y	6	8	7	4	3	5	1	2
d	2	1	1	1	1	2	1	1
d^2	4	1	1	1	1	4	1	1

$$\Sigma d^2 = 14$$

$$r = 1 - \frac{6\Sigma d^2}{n(n^2-1)} = 1 - \frac{6(14)}{8(64-1)}$$

$$= 1 - \frac{84}{504} = 1 - 0.1667$$

$$r = \underline{0.8333}$$

EXAMPLE 5.3.3 Investigate the relationship between the sales of two rival manufacturing companies over the past 12 months by finding the rank correlation coefficient.

Month		J	F	M	A	M	J	J	A	S	O	N	D
Sales (£1000)	A	60	62	74	79	86	94	102	98	100	84	74	78
	B	58	44	45	62	58	65	69	58	67	69	59	64

SOLUTION Ranking the sales gives:

Month	J	F	M	A	M	J	J	A	S	O	N	D
Co. A	12	11	9.5	7	5	4	1	3	2	6	9.5	8
Co. B	9	12	11	6	9	4	1.5	9	3	1.5	7	5
d	3	1	1.5	1	4	0	0.5	6	1	4.5	2.5	3
d^2	9	1	2.25	1	16	0	0.25	36	1	20.25	6.25	9

$$\Sigma d^2 = 102$$

$$r = 1 - \frac{6\Sigma d^2}{n(n^2-1)} = 1 - \frac{6(102)}{12(144-1)}$$

$$= 1 - \frac{612}{1716} = 1 - 0.3566$$

$$r = \underline{0.6434} \quad \text{(i.e. some degree of positive correlation)}$$

EXERCISES ON SECTIONS 5.1 TO 5.3

1. Draw a scatter diagram of each of the sets of values given below, and calculate the PMCC in each case.

   (a)

x	5	6	7	8	9
y	2	5	8	11	14

   (b)

x	2	4	6	8	10	12
y	7	6	5	4	3	2

   (c)

x	1	3	5	7	9	11	13
y	11	7	7	13	8	5	12

2. The following table gives the percentage unemployment figures for males and females in eight regions. Draw a scatter diagram of these data and calculate the PMCC.

Region		London	SE	SW	Midl.	N	NW	Scot.	Wales
Unemployment (%)	Male	7.1	7.6	8.9	8.8	12.3	11.8	10.9	12.6
	Female	6.2	6.5	8.4	7.3	10.2	12.0	10.1	11.3

3. In a job evaluation exercise an assessor ranks nine jobs in order of increasing health risk. The same jobs have also been ranked in decreasing order on the basis of the number of applicants attracted per advertised post.

Job	A	B	C	D	E	F	G	H	I
Health risk	1	2	3	4	5	6	7	8	9
Applicants	3	1	4	2	6	7	5	9	8

Calculate the rank correlation coefficient for this information.

4. The table below gives the shorthand and typing speeds of a sample of 10 secretaries.

Secretary	1	2	3	4	5	6	7	8	9	10
Speed (words/min)										
Typing	40	42	45	45	48	52	54	55	59	60
Shorthand	95	82	96	94	105	96	115	125	96	116

Investigate the degree of correlation between the two skills by calculating (a) the PMCC, and (b) the rank correlation coefficient.

5.4 SIGNIFICANCE OF THE CORRELATION COEFFICIENT

If two variables *are* correlated then the value of r is significantly different from zero. To examine this we could use the t-test, where the value of t is obtained by

$$t = \frac{r\sqrt{n-2}}{\sqrt{1-r^2}}$$

with $(n-2)$ degrees of freedom.

EXAMPLE 5.4.1 Ten pairs of values of the variables x and y gave a correlation coefficient of $r = 0.69$.

Does this value indicate a significant degree of linear correlation between x and y at the 5% level?

SOLUTION We have $r = 0.69$, $n = 10$. We test the assumption that the correlation coefficient is 'near' zero. Now

$$t = \frac{r\sqrt{n-2}}{\sqrt{1-r^2}} = \frac{0.69\sqrt{10-2}}{\sqrt{1-(0.69)^2}}$$

$$= \frac{0.69\sqrt{8}}{\sqrt{0.5239}}$$

$$t = 2.696$$

A significant value of t at the 5% level with $\nu = n - 2 = 10 - 2 = 8$ is found in the tables to be 2.306. Our value of t is larger than this and is therefore significant. The value of $r(= 0.69)$ is thus significantly different from zero. There is therefore a significantly degree of linear correlation between the two variables.

EXAMPLE 5.4.2 Records were kept of the number of personal calls made by eight sales representatives and the total amount of goods sold over a period of three weeks.

Average no. of calls per day	1.9	2.6	3.4	2.9	4.8	3.6	2.8	2.0
Total sales (in £100)	26	30	48	27	48	40	32	21

Calculate the PMCC of this information. How would you interpret your result?

SOLUTION The PMCC is calculated in the table below.

x	y	x^2	y^2	xy
1.9	26	3.61	676	49.4
2.6	30	6.76	900	78
3.4	48	11.56	2304	163.2
2.9	27	8.41	729	78.3
4.8	48	23.04	2304	230.4
3.6	40	12.96	1600	144
2.8	32	7.84	1024	89.6
2.0	21	4.0	441	42
$\Sigma x = 24$	$\Sigma y = 272$	$\Sigma x^2 = 78.18$	$\Sigma y^2 = 9978$	$\Sigma xy = 874.9$

$$\bar{x} = \frac{\Sigma x}{n} = \frac{24}{8} = 3, \qquad \bar{y} = \frac{\Sigma y}{n} = \frac{272}{8} = 34$$

$$r = \frac{\Sigma xy - n\bar{x}\bar{y}}{\sqrt{[\Sigma x^2 - n\bar{x}^2][\Sigma y^2 - n\bar{y}^2]}}$$

$$= \frac{874.9 - 8(3)(34)}{\sqrt{[78.18 - 8(3)^2][9978 - 8(34)^2]}}$$

$$= \frac{58.9}{\sqrt{(6.18)(730)}} = \frac{58.9}{\sqrt{4511.4}}$$

$$r = 0.8769$$

Now, to test whether this indicates a significant correlation we use the t-test, i.e.

$$t = \frac{r\sqrt{n-2}}{\sqrt{1-r^2}} = \frac{0.8769\sqrt{8-2}}{\sqrt{1-(0.8769)^2}} = \frac{2.14796}{0.48067}$$

$$t = 4.469$$

A significant value of t at 5% level with $\nu = n - 2 = 6$ is given by 2.447. Therefore our value of $t(= 4.468)$ is significant, and the value of $r(= 0.8769)$

CORRELATION AND REGRESSION

is significantly different from zero. There is thus a significant degree of linear correlation between the two variables. We conclude that a linear relationship exists between the average number of calls made and the total amount of goods sold by each sales representative.

5.5 REGRESSION

When it is found that a significant degree of correlation exists between x and y (i.e. r is 'near' $+1$ or -1) then these variables are connected by a linear relationship. Thus we are able to calculate the equation of the *regression line* of y on x as given below.

$$y = a + bx$$
$$\text{where} \quad b = \frac{\Sigma xy - n\bar{x}\bar{y}}{\Sigma x^2 - n\bar{x}^2}$$
$$\text{and} \quad a = \bar{y} - b\bar{x}$$

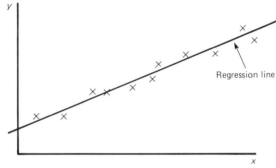

EXAMPLE 5.5.1 The following table shows the staff turnover (the number of leavers as a percentage of the total workforce) for the last seven years.

Year	1976	1977	1978	1979	1980	1981	1982
Turnover	4.6	4.4	4.7	5.1	5.2	5.6	5.8

Draw a scatter diagram of these data and draw the 'best' straight line through the points. Hence estimate the staff turnover in 1983.

SOLUTION

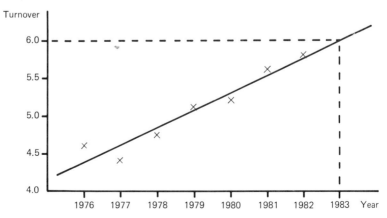

From the diagram we see that the estimate of staff turnover in 1983 = <u>6.0</u>.

EXAMPLE 5.5.2 Assuming that a linear relationship exists, find the equation of the regression line of y on x given the values below.

x	5	6	6	9	10	12
y	2	3	4	4	5	6

Hence estimate a value of y when $x = 15$.

SOLUTION The calculations are shown in the table below.

x	y	x^2	xy
5	2	25	10
6	3	36	18
6	4	36	24
9	4	81	36
10	5	100	50
12	6	144	72
$\Sigma x = 48$	$\Sigma y = 24$	$\Sigma x^2 = 422$	$\Sigma xy = 210$

The equation of the regression line of y on x is

$$y = a + bx$$

where

$$b = \frac{\Sigma xy - n\bar{x}\bar{y}}{\Sigma x^2 - n\bar{x}^2}$$

now

$$\bar{x} = \frac{\Sigma x}{n} = \frac{48}{6} = 8, \qquad \bar{y} = \frac{\Sigma y}{n} = \frac{24}{6} = 4$$

$$b = \frac{210 - 6(8)(4)}{422 - 6(8)^2} = \frac{18}{38} = 0.4737$$

Also

$$a = \bar{y} - b\bar{x} = 4 - (0.4737)(8) = 0.2104$$

Therefore, the regression line equation is

$$\underline{y = 0.21 + 0.47x}$$

Hence when $x = 15$, $y = 0.21 + 0.47(15) = 7.26$.

So when $x = 15$ the best estimate is $\underline{y = 7}$.

EXAMPLE 5.5.3 Find the degree of correlation between the Bank of England base lending rate and the dollar exchange rate taken over the past six months.

Month	Jan	Feb	Mar	Apr	May	June
Base rate (%) (as on 1st of each month)	14	14	13.5	12.5	12	12
Average exchange rate ($)	1.90	1.91	1.86	1.84	1.82	1.83

Hence find an equation relating the two variables and forecast the average exchange rate in a month when the base lending rate is reduced to 11%.

SOLUTION We calculate the PMCC as shown.

x	y	x^2	y^2	xy
14	1.90	196	3.61	26.6
14	1.91	196	3.6481	26.74
13.5	1.86	182.25	3.4596	25.11
12.5	1.84	156.25	3.3856	23
12	1.82	144	3.3124	21.84
12	1.83	144	3.3489	21.96
$\Sigma x = 78$	$\Sigma y = 11.16$	$\Sigma x^2 = 1018.5$	$\Sigma y^2 = 20.7646$	$\Sigma xy = 145.25$

$$\bar{x} = \frac{\Sigma x}{n} = \frac{78}{6} = 13, \qquad \bar{y} = \frac{\Sigma y}{n} = \frac{11.16}{6} = 1.86$$

$$r = \frac{\Sigma xy - n\bar{x}\bar{y}}{\sqrt{[\Sigma x^2 - n\bar{x}^2][\Sigma y^2 - n\bar{y}^2]}}$$

$$= \frac{145.25 - 6(13)(1.86)}{\sqrt{[1018.5 - 6(13)^2][20.7646 - 6(1.86)^2]}}$$

$$= \frac{0.17}{\sqrt{(4.5)(0.007)}} = \frac{0.17}{\sqrt{0.0315}} = \frac{0.17}{0.177\,48}$$

$$r = 0.9578$$

Now r is 'near' $+1$ and therefore indicates a significant linear correlation. To be precise, we can test the above statement by the t-test (see Section 5.4):

$$t = \frac{r\sqrt{n-2}}{\sqrt{1-r^2}} = \frac{0.9578\sqrt{6-2}}{\sqrt{1-(0.9578)^2}} = 6.664$$

From tables, a significant value of t at the 5% level with $\nu = 4$ is 2.776. Therefore, our value of $t(= 6.664)$ is significant, and there is a significant correlation between the two variables.

Hence we can find the equation of the regression line of y on x.

$$y = a + bx \quad \text{least square equation.}$$

where

$$b = \frac{\Sigma xy - n\bar{x}\bar{y}}{\Sigma x^2 - n\bar{x}^2} = \frac{145.25 - 6(13)(1.86)}{1018.5 - 6(13)^2} \qquad 91435$$

$$= \frac{0.17}{4.5} = 0.037\,78$$

and

$$a = \bar{y} - b\bar{x} = 1.86 - (0.037\,78)(13)$$

$$= 1.369$$

So the regression equation is:

$$y = 1.37 + 0.038x$$

When $x = 11$, $y = 1.37 + 0.038(11) = 1.788$.

Therefore, when the lending rate is 11% we predict an exchange rate of $1.79 per £1.

EXAMPLE 5.5.4 The results of two tests taken by nine candidates are tabulated below.

Candidate	A	B	C	D	E	F	G	H	I
Spatial ability	14	6	12	15	7	9	11	8	17
Lateral thinking	19	12	7	9	14	10	16	13	8

Investigate the association between the two sets of results and estimate the score in the lateral thinking test of a tenth candidate who obtained 10 in the test of spatial ability.

SOLUTION Association can be investigated by the PMCC.

x	y	x^2	y^2	xy
14	19	196	361	266
6	12	36	144	72
12	7	144	49	84
15	9	225	81	135
7	14	49	196	98
9	10	81	100	90
11	16	121	256	176
8	13	64	169	104
17	8	289	64	136
$\Sigma x = 99$	$\Sigma y = 108$	$\Sigma x^2 = 1205$	$\Sigma y^2 = 1420$	$\Sigma xy = 1161$

$$\bar{x} = \frac{\Sigma x}{n} = \frac{99}{9} = 11, \qquad \bar{y} = \frac{\Sigma y}{n} = \frac{108}{9} = 12$$

$$r = \frac{\Sigma xy - n\bar{x}\bar{y}}{\sqrt{[\Sigma x^2 - n\bar{x}^2][\Sigma y^2 - n\bar{y}^2]}}$$

$$= \frac{1161 - 9(11)(12)}{\sqrt{[1205 - 9(11)^2][1420 - 9(12)^2]}}$$

$$= \frac{-27}{\sqrt{(116)(124)}} = \frac{-27}{\sqrt{14\,384}}$$

$$r = -0.225$$

Now r is 'near' zero, indicating no significant correlation between the two test scores. Consequently, it would be of little value to obtain the regression equation relating these variables. Furthermore, we are unable to forecast a tenth candidate's score by using this approach.

Formally, to test whether r is 'near' zero we use the t-test:

$$t = \frac{r\sqrt{n-2}}{\sqrt{1-r^2}} = \frac{-0.225\sqrt{7}}{\sqrt{1-(0.225)^2}}$$

$$= \frac{-0.5953}{0.9744} = -0.61$$

From tables, a significant value of t at the 5% level with $v = 7$ is 2.365. Therefore our value is not significant, and the value of r does not indicate a significant linear correlation between the two test scores.

EXERCISES ON SECTIONS 5.4 AND 5.5

1. From the following sets of data examine whether a linear relationship exists between x and y. Where appropriate, find the equation of the regression line of y on x. Hence estimate the value of y corresponding to the value of x given in brackets.

   (a)

x	2	4	6	8	10	($x = 12$)
y	3	4	6	7	10	

   (b)

x	1	3	5	7	9	11	($x = 10$)
y	13	12	10	10	8	7	

   (c)

x	2	4	5	5	6	7	11	($x = 12$)
y	6	9	4	2	7	3	4	

2. The given table shows the advertising expenditure and resulting sales of a company over eight months.

Month	1	2	3	4	5	6	7	8
Advertising expenditure (× £100)	1.9	2.4	2.3	3.7	2.9	2.6	3.1	2.7
Sales (× £10 000)	10	12	13	17	14	13	15	14

   Calculate the degree of correlation between sales and advertising expenditure.

   Furthermore find the equation of the regression line in order to forecast the sales in a given month in which £400 is spent on advertising.

3. The total value of goods exported by a company over the past six years is given in the table below. Forecast the value of goods exported during 1983.

Year	1977	1978	1979	1980	1981	1982
Exports (× £1000)	4.5	3.7	5.2	5.7	6.8	7.1

4. The life of a machine part depends upon the speed at which the machine operates. The following data give the lifetimes of 10 machine parts operating at different speeds.

Speed (1000 rev/min)	0.9	0.9	1.0	1.1	1.3	1.5	1.5	1.6	1.8	2.4
Lifetime (hours)	152	164	154	146	124	116	120	117	108	109

Estimate the lifetime of a part if the machine operates at 2500 rev/min.

PAST QUESTIONS 5

1. (a) Explain the purpose of calculating a correlation coefficient and give the possible range of values it may take.

   (b) Suppose that it is found, using some statistical test, that a calculated correlation coefficient of $+0.6$ between two variables is not significantly different from zero. What does this mean?

   (c) The following table gives the monthly output, x, and labour cost, y, of a factory.

Monthly output (tons $\times 10^3$)	Labour cost (£ $\times 10^3$)
66	50
74	53
78	59
70	52
81	64
90	85
87	77
85	68

   Required:

   Calculate the correlation coefficient using

   $$r = \frac{n\Sigma xy - \Sigma x \Sigma y}{\sqrt{[n\Sigma x^2 - (\Sigma x)^2][n\Sigma y^2 - (\Sigma y)^2]}}$$

   or some alternative formula that you prefer, and comment on your result. How would your answer be affected if output were measured in tonnes, instead of tons, where 1 ton $= 1.016$ tonne? [ACA]

2. A transport consultant has been commissioned to advise a developing country on its future energy requirements and on road/rail capacity needed to transport the necessary fuels. The consultant found the following statistical information.

Country	U	C	D	A	R	J	B
GNP per capita, in 100 dollars	29	19	15	14	8	5	2
Industrial energy consumption per capita, 10^7 BTU	18	13	8	11	7	3	1

   (a) Calculate the correlation coefficient for the two variables shown.

(b) Use the appropriate statistical method to make an estimate of the country's energy requirements if the GNP increases from the present level of 300 to 800.

(c) Comment on the usefulness of this type of analysis, for the purpose described. [CIT]

3.

Data set	1	2	3	4	5	6	7	8	9	10
Basic mins	17	19	20	20	24	28	29	29	31	39
Variable 1	320	330	400	380	460	530	580	560	600	740
Variable 2	50	62	62	64	68	75	85	89	89	100

(a) (i) Plot two scatter diagrams, one for each variable against basic time and draw a line of best fit for each (by eye or any other method).
(ii) Using the lines of best fit, estimate the basic time for a value of 500 of variable 1 and for a value of 75 for variable 2.

(b) Explain how both variables might be used together to predict a basic time and comment on the effect of the obvious correlation between variable 1 and variable 2. [IMS]

4. (a) Explain fully the following terms:
(i) linear regression,
(ii) correlation analysis,
(iii) scatter diagram.

(b) The personnel department of your company is considering the possibility of assessing applicants by using psychological tests instead of the normal interview procedure. A comparative test of seven applicants has been carried out using both methods and the following results obtained.

Applicant	Ranking by interview	Ranking by tests
A	4	5
B	1	2
C	7	7
D	6	4
E	2	1
F	3	3
G	5	6

Advise the personnel department on the degree of agreement between the two methods. [IIM]

5. (a) Explain the difference in use between regression analysis and correlation analysis.

(b) Calculate the correlation coefficient for the following information which shows the share price of a particular company and the Financial Times Ordinary Share Index compared on randomly selected days. Comment upon the result.

Share price (p)	82	84	89	73	94	83	85	75
FT index	401	418	427	414	486	454	439	451

[IM]

6. A medium sized manufacturing organisation has the following sales analysis for the period 1971–80 inclusive.

Year	Sales (£1 million)
1971	15.3
1972	14.6
1973	16.8
1974	17.3
1975	17.2
1976	20.9
1977	22.3
1978	20.0
1979	23.1
1980	24.5

You are required to:

(a) Calculate:
   (i) the trend line using the least squares equation,
   (ii) the trend values for 1971 and 1980 and explain the meaning of these results.

(b) Draw a graph of the above data including the trend and project the trend for the year 1981.   [ICMA]

7. (a) Define what is meant by the term correlation and explain exactly what a measure of correlation does measure, indicating the strengths and weaknesses of such a measure.

(b) Briefly explain what is meant by the term 'coefficient of determination'.

(c) The following data was extracted from the 1971 Census for a large city in England.

Ward	Percentage of households with one or more cars	Percentage of families with four or more children
A	27.86	6.70
B	20.27	2.34
C	59.53	1.83
D	21.79	4.22
E	23.25	9.24
F	33.32	3.85
G	24.10	10.33
H	30.13	8.48
I	59.82	2.81
J	35.67	3.25

You are required to:

(i) Explain whether the Pearsonian (product-moment) or rank Spearman coefficient is more useful for this data and why.

(ii) Calculate a Pearsonian (or product-moment) correlation coefficient.

(iii) Describe graphically the general conclusions which can be made from your result in (ii).   [CIPFA]

8. (*a*) Explain the essential difference between the use of the product moment correlation coefficient and the rank correlation coefficient as used in correlation analysis.

(*b*) Three interviewers assess six candidates for a post and rank them as follows:

	Candidates					
	A	B	C	D	E	F
Interviewer 1	1	2	3	4	5	6
Interviewer 2	4	3	1	2	6	5
Interviewer 3	3	1	2	4	6	5

Using the Spearman rank correlation coefficient R_S as the measure of agreement, which pair of interviewers showed the greatest agreement?

[IIM]

9. For purposes of forward planning, a passenger transport organisation has been examining regional variations in the use of buses. Amongst other factors, car ownership levels obviously seem important. Certain investigations produced the following results:

Percentage of households with at least one car	46	56	42	65	59	50	51	45
Bus trips per household per week	10	4	13	5	3	7	9	10

(*a*) Calculate the correlation coefficient for these data.

(*b*) Plot the data and find the linear regression equation. Would there be a better relationship to fit the data?

(*c*) What other factors are likely to cause the regional variations in the use of buses? How would you analyse the effects of these? [CIT]

10. An examination of records supplied by ten firms engaged in a similar type of production provides the following data.

Capital per employee (× £1000)	Output per employee (× £1000)
4.30	1.50
4.00	1.30
0.75	0.50
3.25	0.80
1.25	0.70
1.40	0.60
2.50	1.00
1.50	0.80
2.30	1.00
1.75	0.80

(*a*) Plot the data in the form of a scatter diagram with capital per employee as '*x*' and output per employee as '*y*'.

(*b*) Determine the regression equation for predicting output from capital ($\Sigma x^2 = 65.98$, $\Sigma xy = 23.74$).

(c) If a firm comparable to these firms wished to obtain an output of £1250 per employee, what level of capital is indicated by the regression equation? [IIM]

11. A manufacturing company has ten machines of similar type. It is investigating the relationship between the weekly cost of maintenance of these machines and their age. Figures for October 1981 were as follows.

Machine	1	2	3	4	5	6	7	8	9	10
Age (months), X	5	10	15	20	30	30	30	50	50	60
Maintenance (£), Y	190	240	250	300	310	335	300	330	350	395

$\Sigma X^2 = 12\,050$ $\Sigma XY = 99\,150$

(a) Draw a scatter diagram of the data.

(b) Find the least squares regression of maintenance cost on age.

(c) Explain the practical meaning of the regression equation to a colleague who is not trained in statistics.

(d) Predict the maintenance cost for a machine of this type which is 40 months old. Why might this estimate be inaccurate? [ICMA]

12. The trade union have complained that their earnings potential has been reduced by the amount of time lost due to machine break-down and material shortages. You are required to analyse the following data and comment on the validity of the trade union claim.

Week no.	Average earnings (£)	Lost time (%)
1	75	15
2	90	10
3	100	10
4	77	12
5	108	4
6	85	17
7	70	23
8	90	14
9	94	10
10	79	21
11	98	5
12	77	20

[IPM]

13. A sample of eight employees is taken from the production department of a light engineering factory. The data below relates to the number of weeks' experience in the wiring of components, and the number of components which were rejected as unsatisfactory last week.

Employee	A	B	C	D	E	F	G	H
Weeks of experience (X)	4	5	7	9	10	11	12	14
No. of rejects (Y)	21	22	15	18	14	14	11	13

$\Sigma X = 72$ $\Sigma Y = 128$ $\Sigma XY = 1069$ $\Sigma X^2 = 732$ $\Sigma Y^2 = 2156$

(a) Draw a scatter diagram of the data.

(b) Calculate a coefficient of correlation for these data and interpret its value.

(c) Find the least squares regression equation of rejects on experience. Predict the number of rejects you would expect from an employee with one week of experience. [ICMA]

14. The cost, c, of providing a particular service is made up of two parts, a fixed cost, f, and a variable cost per unit serviced, v. The expected relationship between these values is $c = f + nv$ where n is the number of units serviced. From the ten actual values of c and n given below, estimate values for f and v. Calculate and comment upon the correlation coefficient.

Cost (£c)	120	230	250	400	390	580	580	670	670	800
No. of units (n)	5	8	10	18	19	25	27	30	31	35

[IMS]

15. (a) Describe how regression analysis is of help in market research.

(b) Calculate the equation of the regression line for the following information showing the paired relationship between weekly units of production (y) and man-days lost through absenteeism (x).

y	48	41	23	39	18	27	30	35
x	5	8	17	9	19	14	13	11

[IM]

16.

Age in years (x)	18	20	21	27	23	34	24	42	38	44
Length of training required in days (y)	8	5	6	8	7	11	8	10	6	8

The length of training (y) required by operatives to be able to perform a routine task is shown in the above table, along with the age (x) of the trainees. Outline the procedure to determine whether a relationship exists, close or otherwise, between the two variables. [ITD]

17. There have been 10 deaths amongst the pensioners of a pension scheme over the last year. Set out below is a table showing their ages at death and the age at which each one actually retired.

Age at death	Age at retirement
88	60
82	60
80	63
79	63
77	63
73	65
73	61
70	64
70	65
68	66

Calculate the coefficient of correlation between age at retirement and age at death assuming all ages are exact years.

Comment on the significance of the resulting figure. [CII]

18. Worker	A	B	C	D	E	F	G	H	I	J
Test % (X)	85	82	76	70	66	60	60	58	56	55
Performance % (Y)	77	86	90	75	80	70	74	65	65	65

The table above shows how ten workers, designated by the letters A to J scored in a selection test (X) and how their work performances were subsequently rated (Y). Outline the procedure to determine whether a relationship exists, close or otherwise, between the two sets of scores. [ITD]

6 FORECASTING

6.1 TREND

The trend in a time series can be estimated by regression analysis (see Section 5.5) or by graphical techniques.

EXAMPLE 6.1.1 From the information given below estimate the total production in the eleventh month.

Month	1	2	3	4	5	6	7	8	9	10
Production (1000 tonnes)	12.1	12.4	12.5	12.9	13.0	13.2	13.3	13.6	14.0	14.2

SOLUTION

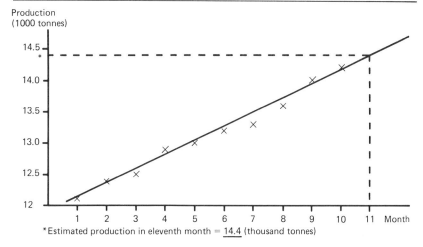

*Estimated production in eleventh month = __14.4__ (thousand tonnes)

EXAMPLE 6.1.2 The following table gives the company sales of a certain product over a 12-year period. Forecast the sales in 1984.

Year	1972	1973	1974	1975	1976	1977	1978	1979	1980	1981	1982	1983
Sales (£10 000)	30	24	25	20	23	21	24	24	31	40	51	64

SOLUTION

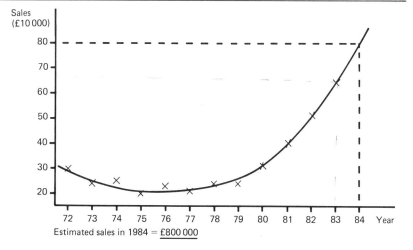

Estimated sales in 1984 = __£800 000__

6.2 MOVING AVERAGES

Moving averages can be used to smooth out a time series in order to isolate the *trend*.

EXAMPLE 6.2.1

Year	1974	1975	1976	1977	1978	1979	1980	1981	1982	1983	1984
Profit (× £1000)	20	23	32	51	39	22	25	36	58	70	52

Isolate the trend in the time series of annual profit given above by finding:

(a) three-point moving averages,

(b) five-point moving averages.

SOLUTION The time series is illustrated in the graph below.

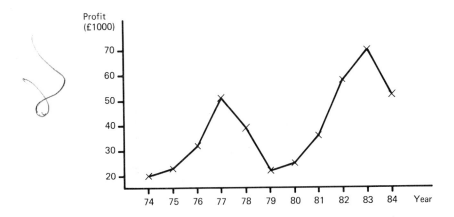

The moving averages are calculated as follows.

Year	Profit	Three-point moving average	Five-point moving average
1974	20		
1975	23	25.00	
1976	32	35.33	33.0
1977	51	40.67	33.4
1978	39	37.33	33.8
1979	22	28.67	34.6
1980	25	27.67	36.0
1981	36	39.67	42.2
1982	58	54.67	48.2
1983	70	60.00	
1984	52		

The two sets of moving averages are superimposed on to the graph of annual profit over the period 1974–84 shown on the following page.

FORECASTING 113

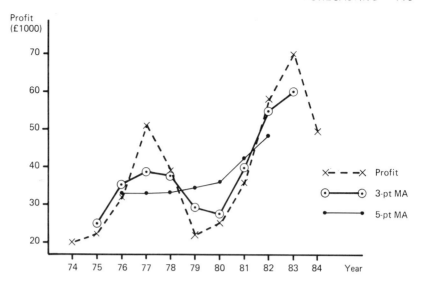

EXAMPLE 6.2.2 The sale of Local Authority houses to existing tenants over a four-year period is tabulated below.

Year	No. of houses sold		
	Jan–Apr	May–Aug	Sept–Dec
1980	43	85	64
1981	64	100	70
1982	79	124	88
1983	91	142	103

Plot these values on to a suitable graph. Superimpose the three-point moving averages in order to highlight the trend in this time series.

SOLUTION The three-point moving averages are calculated in the following table.

Period		No. of sales	Three-point moving average
1980	Jan–Apr	43	
	May–Aug	85	64
	Sept–Dec	64	71
1981	Jan–Apr	64	76
	May–Aug	100	78
	Sept–Dec	70	83
1982	Jan–Apr	79	91
	May–Aug	124	97
	Sept–Dec	88	101
1983	Jan–Apr	91	107
	May–Aug	142	112
	Sept–Dec	103	

The diagram on the next page illustrates the sale of council houses together with the trend (three-point moving averages).

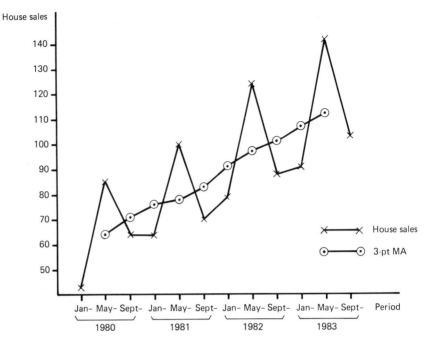

EXAMPLE 6.2.3 The quarterly production figures for a large manufacturing company are given below. Use four-point moving averages to isolate the trend. Plot the centred moving averages on to a graph.

	Total production (thousand units)				
	1980	1981	1982	1983	1984
1st quarter	–	152	148	138	134
2nd quarter	–	154	150	140	–
3rd quarter	138	130	131	122	–
4th quarter	148	145	140	138	–

SOLUTION

Quarter		Production	Four-point moving average	Centred moving average
1980	3rd	138		
	4th	148		
			148	147
1981	1st	152	146	145.625
	2nd	154	145.25	144.75
	3rd	130	144.25	143.75
	4th	145	143.25	143.375
1982	1st	148	143.5	142.875
	2nd	150	142.25	141
	3rd	131	139.75	138.5
	4th	140	137.25	136.125
1983	1st	138	135	134.75
	2nd	140	134.5	134
	3rd	122	133.5	
	4th	138		
1984	1st	134		

The centred moving averages illustrate the trend on the graph below.

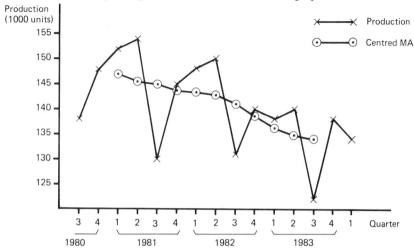

EXAMPLE 6.2.4 The daily revenue of a department store is tabulated below over a three-week period. Find the six-point moving averages and plot the centred values on to a graph. Use your graph to forecast the *trend* values during the fourth week.

	Daily revenue (× £100)					
	Mon	Tue	Wed	Thur	Fri	Sat
Week 1	28	25	29	29	36	54
Week 2	30	25	30	34	37	56
Week 3	31	28	33	34	40	59

SOLUTION The six-point moving averages and centred averages are shown in the following table.

Day	Revenue (× £100)	Six-point moving average	Centred moving average
Mon	28		
Tue	25		
Wed	29		
		33.5	
Thur	29		33.667
		33.833	
Fri	36		33.833
		33.833	
Sat	54		33.917
		34	
Mon	30		34.417
		34.833	
Tue	25		34.917
		35	
Wed	30		35.167
		35.333	
Thur	34		35.417
		35.5	
Fri	37		35.75
		36	
Sat	56		36.25
		36.5	
Mon	31		36.5
		36.5	
Tue	28		36.75
		37	
Wed	33		37.25
		37.5	
Thur	34		
Fri	40		
Sat	59		

The centred moving averages are plotted on the graph below.

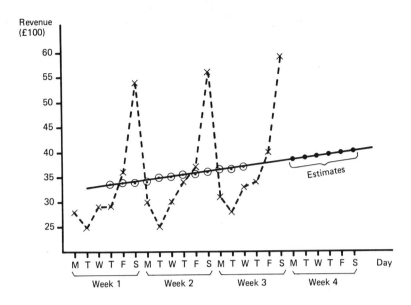

By drawing the 'best' straight line (regression line) through the centred moving averages we obtain the following estimates of the trend in week 4.

Day	Mon	Tue	Wed	Thur	Fri	Sat
Trend	38.5	38.9	39.2	39.6	39.9	40.2

[*Note*: These are estimates of the *trend*, and not estimates of the revenue in week 4.]

6.3 SEASONAL VARIATIONS

Seasonal variations can be combined with the trend in order to obtain a simple forecasting model, i.e. $X = T + S$

(the *additive model*) where X = value to be forecasted, T = estimated trend, and S = estimated seasonal variation.

[*Note*: In general, this model is written as $X = T + S + R$ where R = residual variation. For the purposes of the following examples, we will consider $R = 0$.]

EXAMPLE 6.3.1 Calculate the deviations between the monthly sales figures and corresponding moving averages tabulated below.

Month	Jan	Feb	Mar	Apr	May	June	July	Aug	Sept
Sales (× £1000)	29	34	38	46	37	34	34	38	48
Moving averages	33.1	34.7	34.9	35.2	36.7	37.5	38.7	40.1	40.6

SOLUTION The deviations are calculated in the following table.

Month	Sales (X)	Moving averages (T)	Deviations ($S = X - T$)
Jan	29	33.1	−4.1
Feb	34	34.7	−0.7
Mar	38	34.9	3.1
Apr	46	35.2	10.8
May	37	36.7	0.3
June	34	37.5	−3.5
July	34	38.7	−4.7
Aug	38	40.1	−2.1
Sept	48	40.6	7.4

EXAMPLE 6.3.2 The table below shows the staff turnover (the number of leavers as a percentage of the total workforce) experienced by a company over a four-year period.

Year	Quarter			
	1st	2nd	3rd	4th
1	3.1	2.6	4.8	4.3
2	3.7	2.8	5.2	4.7
3	4.1	3.4	5.8	4.9
4	4.5	4.0	6.0	5.3

By finding average deviations from the trend, estimate the seasonal variations for each of the four quarters.

SOLUTION The deviations from the trend are calculated below.

Quarter	Turnover (X)	Four-point moving average	Centred moving average (T)	Deviations
1st	3.1			
2nd	2.6			
		3.7		
3rd	4.8		3.775	1.025
		3.85		
4th	4.3		3.875	0.425
		3.9		
1st	3.7		3.95	−0.25
		4.0		
2nd	2.8		4.05	−1.25
		4.1		
3rd	5.2		4.15	1.05
		4.2		
4th	4.7		4.275	0.425
		4.35		
1st	4.1		4.425	−0.325
		4.5		
2nd	3.4		4.525	−1.125
		4.55		
3rd	5.8		4.6	1.2
		4.65		
4th	4.9		4.725	0.175
		4.8		
1st	4.5		4.825	−0.325
		4.85		
2nd	4.0		4.9	−0.9
		4.95		
3rd	6.0			
4th	5.3			

The deviations in each quarter are summarised below.

Year	Quarter				Total of averages
	1st	2nd	3rd	4th	
1	–	–	1.025	0.425	
2	−0.25	−1.25	1.05	0.425	
3	−0.325	−1.125	1.2	0.175	
4	−0.325	−0.9	–	–	
Average deviations	−0.3	−1.092	1.092	0.342	0.042
Corrected deviations (−0.042/4)	−0.310	−1.102	1.081	0.331	

Therefore, the estimated seasonal variations are:

1st quarter: −0.31 2nd quarter: −1.10
3rd quarter: 1.08 4th quarter: 0.33

EXAMPLE 6.3.3 The table below shows the total expenditure on part-time teaching by a Local Education Authority over a four-year period.

Term	Expenditure (× £1000)			
	1980–81	1981–82	1982–83	1983–84
Autumn	156	150	141	132
Spring	102	96	93	84
Summer	66	63	54	51

(a) Isolate the trend by finding the three-point moving averages. Estimate the seasonal variations for the three terms.

(b) Forecast the total expenditure during each term in the academic year 1984–85.

SOLUTION (a) The moving averages and corresponding deviations are tabulated below.

Term	Expenditure (X)	Three-point moving average (T)	Deviations (S)
1980–81: Autumn	156		
Spring	102	108	−6
Summer	66	106	−40
1981–82: Autumn	150	104	46
Spring	96	103	−7
Summer	63	100	−37
1982–83: Autumn	141	99	42
Spring	93	96	−3
Summer	54	93	−39
1983–84: Autumn	132	90	42
Spring	84	89	−5
Summer	51		

The deviations are used to estimate the seasonal variations in the following table.

Year	Autumn	Spring	Summer	Total averages
1980-81	–	−6	−40	
1981-82	46	−7	−37	
1982-83	42	−3	−39	
1983-84	42	−5		
Average deviations	43.333	−5.25	−38.667	−0.584
Corrected deviations (+0.584/3)	43.528	−5.055	−38.472	

Therefore the forecasted seasonal variations (S) are:

Autumn: <u>43.53</u>   Spring: <u>−5.06</u>   Summer: <u>−38.47</u>

(b) The trend is forecasted in the graph below by fitting the 'best' straight line to the moving averages.

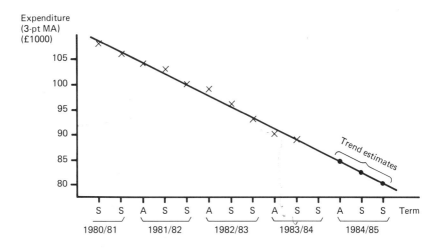

From the graph the trend estimates (T) in 1984-85 are:

Autumn: <u>84.2</u>   Spring: <u>82</u>   Summer: <u>79.8</u>

Combining the trend and seasonal variations $(X = T + S)$ we obtain forecasts of the expenditure during 1984-85:

Autumn: 84.2 + 43.53 = 127.73
Spring: 82 − 5.06   = 76.94
Summer: 79.8 − 38.47 = 41.33

Thus, the estimated expenditures during the academic year 1984-85 are:

Autumn: <u>£128 000</u>   Spring: <u>£77 000</u>   Summer: <u>£41 000</u>

EXAMPLE 6.3.4 From the table of quarterly magazine sales given below, forecast the sales in each quarter of 1984. (The figures are given in thousands of copies.)

	1980	1981	1982	1983
1st quarter	112	132	144	160
2nd quarter	145	161	169	181
3rd quarter	164	180	196	208
4th quarter	123	135	143	155

SOLUTION

Quarter	Sales (X)	Four-point moving average	Centred moving average (T)	Deviations (X−T)
1980: 1st	112			
2nd	145			
		136		
3rd	164		138.5	25.5
		141		
4th	123		143	−20
		145		
1981: 1st	132		147	−15
		149		
2nd	161		150.5	10.5
		152		
3rd	180		153.5	26.5
		155		
4th	135		156	−21
		157		
1982: 1st	144		159	−15
		161		
2nd	169		162	7
		163		
3rd	196		164	32
		167		
4th	143		168.5	−25.5
		170		
1983: 1st	160		171	−11
		173		
2nd	181		174.5	6.5
		176		
3rd	208			
4th	155			

The seasonal variations are estimated in the table below.

Year	1st	2nd	3rd	4th	Total averages
1980	−	−	25.5	−20	
1981	−15	10.5	26.5	−21	
1982	−15	7	32	−25.5	
1983	−11	6.5	−	−	
Average deviations	−13.667	8	28	−22.167	0.166
Corrected deviations (−0.166/4)	−13.709	7.958	27.958	−22.209	

Therefore the seasonal variations (S) are:

  1st quarter: −13.71   2nd quarter: 7.96
  3rd quarter: 27.96   4th quarter: −22.21

The trend is shown in the graph below.

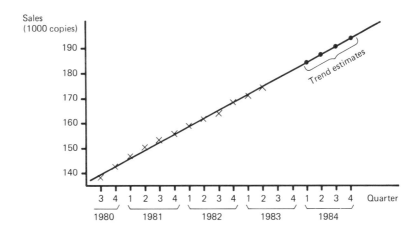

From the graph the trend estimates for 1984 are:

  1st quarter: <u>184</u>   2nd quarter: <u>187</u>
  3rd quarter: <u>190</u>   4th quarter: <u>193</u>

Combining the trend and seasonal variations we have:

  1st quarter: $184 - 13.71 = 170.29$
  2nd quarter: $187 + 7.96 = 194.96$
  3rd quarter: $190 + 27.96 = 217.96$
  4th quarter: $193 - 22.21 = 170.79$

Therefore, the forecasted sales in 1984 are:

Quarter	1st	2nd	3rd	4th
Sales	£170 000	£195 000	£218 000	£171 000

EXERCISES ON SECTIONS 6.1 TO 6.3

1. Illustrate the data given below.

Year	1970	1971	1972	1973	1974	1975	1976	1977	1978	1979	1980	1981	1982
Income (£ millions)	2.6	2.7	2.9	2.8	3.1	3.6	3.4	3.5	3.9	3.9	4.3	4.6	4.6

Using the same graph, illustrate the moving averages obtained from (*a*) three points, and (*b*) seven points.

2. Plot the data given below on to a suitable graph. Find the six-point moving averages and plot the centred moving averages on to the same diagram.

Year	1973	1974	1975	1976	1977	1978	1979	1980	1981	1982	1983	1984
Sales (× £1000)	24	29	27	28	32	31	36	37	37	39	42	41

Hence estimate the 'trend' in 1985.

3. The table below gives the total export orders for a company over four-monthly periods during the years 1980-83. Evaluate three-point moving averages in order to isolate the trend. By finding seasonal variations forecast the total export orders during the three periods of 1984.

Year	Total exports (£100 000)		
	Jan–Apr	May–Aug	Sept–Dec
1980	3.4	4.5	3.8
1981	4.0	4.8	4.1
1982	4.3	5.7	4.7
1983	4.9	5.7	5.0

4. By finding the trend and seasonal variations, forecast quarterly values in the fourth year for the time series given below.

(a)

	1st quarter	2nd quarter	3rd quarter	4th quarter
Year 1	12	24	13	13
Year 2	20	33	19	20
Year 3	27	41	25	26

(b)

	1st quarter	2nd quarter	3rd quarter	4th quarter
Year 1	43	36	40	47
Year 2	39	34	34	43
Year 3	33	28	30	39

5. The table below shows the total withdrawals from savings accounts in a national building society over the years 1980-83.

Period	Total withdrawals (£ millions)			
	1980	1981	1982	1983
Jan–Mar	204	252	308	352
Apr–June	296	348	420	480
July–Sept	288	386	422	470
Oct–Dec	202	256	320	392

Estimate the withdrawals during the corresponding periods in 1984.

6.4 THE MULTIPLICATIVE MODEL

When forecasting from certain time series a multiplicative model is used in preference to an additive model, i.e.

$$X = T \cdot S$$

where T = trend and S = seasonal index.

[*Note*: In general, the multiplicative model is $X = T \cdot S \cdot R$ where R = residual index. For the purpose of isolating T and S, we will ignore the effect of R, i.e. we assume $R = 1$.]

FORECASTING 123

EXAMPLE 6.4.1 State whether an additive or multiplicative model would be more appropriate in approximating the time series given below.

(a)
Period	1	2	3	4	5	6	7	8	9	10	11	12
X	10	30	35	16	19	39	42	24	25	49	51	33

(b)
Period	1	2	3	4	5	6	7	8	9	10	11	12
X	12	33	42	16	23	54	64	25	36	72	85	31

SOLUTION (a)

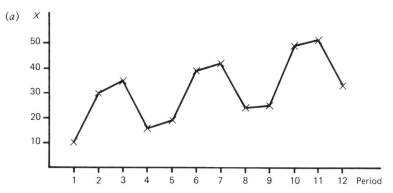

The additive model can be used in this case, i.e. $\underline{X = T + S}$.

(b)

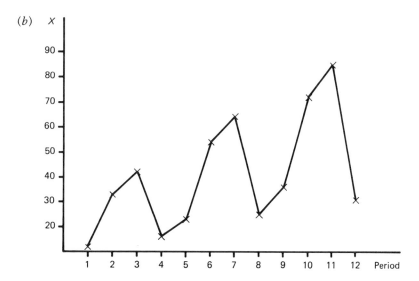

The multiplicative model is more appropriate in this case, i.e. $\underline{X = T \cdot S}$.

EXAMPLE 6.4.2 Assuming a multiplicative model, calculate the seasonal indices given the sequence of values together with the moving averages shown below.

X	16	31	39	20	44	57	26	61	71	32	74	84
Moving averages			23	26	30	33	37	41	44	48	51	55	58	62

SOLUTION [*Note:* We have $X = T \cdot S$.]

$$\therefore \quad S = \frac{X}{T} \quad \text{(the seasonal index)}$$

X	Moving averages (T)	Seasonal indices (S)
16	23	0.696
31	26	1.192
39	30	1.3
20	33	0.606
44	37	1.189
57	41	1.390
26	44	0.591
61	48	1.271
71	51	1.392
32	55	0.582
74	58	1.276
84	62	1.355

EXAMPLE 6.4.3 The table below gives the number of new consumers registered by the Consumer Services Department in a regional gas-board headquarters.

Year	Jan–Apr	May–Aug	Sept–Dec
1980	–	440	864
1981	448	526	1062
1982	560	702	1298
1983	665	822	1541

By using a multiplicative model, forecast the number of consumer registrations during the three periods in 1984.

SOLUTION

Period		Consumer registrations (X)	Three-point moving average (T)	$S (= X/T)$
1980	May–Aug	440		
	Sept–Dec	864	584	1.479
1981	Jan–Apr	448	612.67	0.731
	May–Aug	526	678.67	0.775
	Sept–Dec	1062	716	1.483
1982	Jan–Apr	560	774.67	0.723
	May–Aug	702	853.33	0.823
	Sept–Dec	1298	888.33	1.461
1983	Jan–Apr	665	928.33	0.716
	May–Aug	822	1009.33	0.814
	Sept–Dec	1541		

The seasonal indices are shown in the following table.

Year	Jan–Apr	May–Aug	Sept–Dec
1980	–	–	1.479
1981	0.731	0.775	1.483
1982	0.723	0.823	1.461
1983	0.716	0.814	–
Average index (S)	0.723	0.804	1.474

The trend is forecasted in the graph of moving averages shown below.

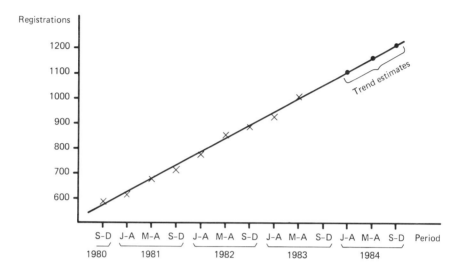

The trend estimates from the graph are <u>1106</u>, <u>1160</u> and <u>1214</u>. Therefore, the forecasts for 1984 are:

Jan–Apr: $1106 \times 0.723 = 799.6$
May–Aug: $1160 \times 0.804 = 932.6$
Sept–Dec: $1214 \times 1.474 = 1789.4$

Thus the estimated numbers of new consumer registrations during 1984 are:

<u>Jan–Apr: 800   May–Aug: 933   Sept–Dec: 1789</u>

EXAMPLE 6.4.4 The number of package tours booked through a travel agency over a three-year period is given in the table below.

	1980	1981	1982
1st quarter	156	152	160
2nd quarter	314	365	423
3rd quarter	301	349	405
4th quarter	112	127	150

Estimate the number of tours booked during each quarter in 1983.

SOLUTION The data are shown in the graph below.

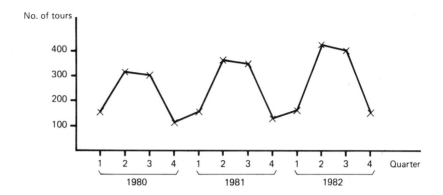

The graph shows that a multiplicative model is most appropriate, i.e. $X = T \cdot S$.

Quarter	No. of tours (X)	Four-point moving average	Centred moving average (T)	Seasonal indices (X/T)
1980: 1st	156			
2nd	314			
		220.75		
3rd	301		220.25	1.367
		219.75		
4th	112		226.125	0.495
		232.5		
1981: 1st	152		238.5	0.637
		244.5		
2nd	365		246.375	1.481
		248.25		
3rd	349		249.25	1.400
		250.25		
4th	127		257.5	0.493
		264.75		
1982: 1st	160		271.75	0.589
		278.75		
2nd	423		281.625	1.502
		284.5		
3rd	405			
4th	150			

The seasonal indices are given below.

Year	1st quarter	2nd quarter	3rd quarter	4th quarter
1980	–	–	1.367	0.495
1981	0.637	1.481	1.400	0.493
1982	0.589	1.502	–	–
Average index (S)	0.613	1.492	1.384	0.494

The trend during 1983 is forecasted from the graph of centred moving averages shown opposite.

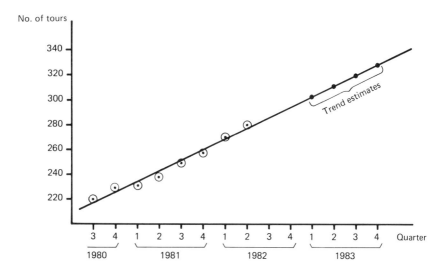

From the graph the trend estimates for the four quarters in 1983 are 305, 314, 323 and 332. Therefore, estimates for 1983 are:

1st quarter: 305 × 0.613 = 186.965
2nd quarter: 314 × 1.492 = 468.488
3rd quarter: 323 × 1.384 = 447.032
4th quarter: 332 × 0.494 = 164.008

Thus the estimated number of bookings during 1983 are:

1st quarter: 187	2nd quarter: 468
3rd quarter: 447	4th quarter: 164

6.5 EXPONENTIAL SMOOTHING

Exponential smoothing is an alternative method of isolating the trend in a time series. We have

$$T_i = \alpha X_i + (1-\alpha)T_{i-1}$$

where α = smoothing constant (between 0 and 1), X_i = actual value in ith period, and T_i = trend in ith period.

EXAMPLE 6.5.1 By using exponential smoothing isolate the trend in the sales figures given below with (a) $\alpha = 0.1$, (b) $\alpha = 0.3$ and (c) $\alpha = 0.5$.

Week	1	2	3	4	5	6	7	8	9	10
Sales (× £1000)	25	19	16	24	40	20	12	28	36	27

SOLUTION

Week (i)	Sales (X_i)	$T_i(\alpha = 0.1)$	$T_i(\alpha = 0.3)$	$T_i(\alpha = 0.5)$
1	25	25	25	25
2	19	24.4	23.2	22
3	16	23.56	21.04	19
4	24	23.60	21.93	21.5
5	40	25.24	27.35	30.75
6	20	24.72	25.14	25.38
7	12	23.45	21.20	18.69
8	28	23.90	23.24	23.34
9	36	25.11	27.07	29.67
10	27	25.30	27.05	28.34

The sales together with the exponentially smoothed values are shown on the graph below.

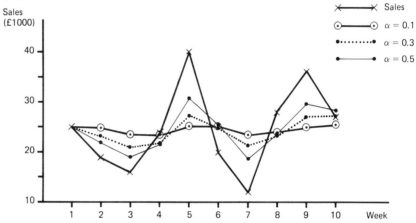

EXAMPLE 6.5.2 Smooth out the time series in the table below by using a smoothing constant of (a) $\alpha = 0.2$, (b) $\alpha = 0.6$. Assume that the average profit prior to 1972 is £42 000.

Year	1972	1973	1974	1975	1976	1977	1978	1979	1980	1981	1982	1983
Profit (× £1000)	41	52	56	43	49	64	57	52	59	68	72	65

SOLUTION

Year	Profit (X_i)	$T_i(\alpha = 0.2)$	$T_i(\alpha = 0.6)$
Pre-1972	–	42.00	42.00
1972	41	41.80	41.40
1973	52	43.84	47.76
1974	56	44.64	52.70
1975	43	44.31	46.88
1976	49	45.25	48.15
1977	64	49.00	57.66
1978	57	50.60	57.26
1979	52	50.88	54.11
1980	59	52.50	57.04
1981	68	55.60	63.62
1982	72	58.88	68.65
1983	65	60.11	66.46

The annual profit and 'trend' values are illustrated in the following graph.

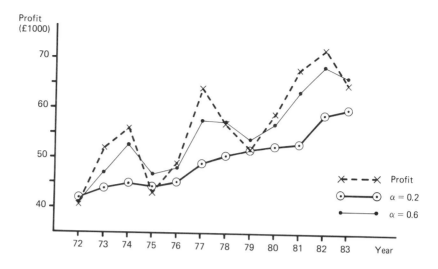

EXERCISES ON SECTIONS 6.4 AND 6.5

1. Sketch the graphs of the following time series and state in each case whether it would be appropriate to use an additive or a multiplicative model when forecasting.

(a)

Period	1	2	3	4	5	6	7	8	9	10	11	12
X	19	29	27	15	23	37	35	18	28	46	43	20

(b)

Period	1	2	3	4	5	6	7	8	9	10	11	12
X	124	133	113	100	135	142	124	112	146	157	138	129

(c)

Period	1	2	3	4	5	6	7	8	9	10	11	12
X	38	77	41	34	65	30	34	53	26	24	43	22

2. The average number of unemployed (excluding school leavers) in a small town during the period 1979–82 is shown in the table below. Use a suitable model to forecast the unemployment figures during 1983.

Period	1979	1980	1981	1982
Jan–Apr	328	432	546	640
May–Aug	216	280	320	374
Sept–Dec	370	458	561	634

3. Use the information tabulated below to estimate the number of houses under construction during the four quarters in 1984.

Period	No. of houses under construction			
	1980	1981	1982	1983
Jan–Mar	330	405	485	540
Apr–June	620	735	845	985
July–Sept	675	790	915	1065
Oct–Dec	410	500	595	665

4. Use exponential smoothing with (a) $\alpha = 0.1$, (b) $\alpha = 0.4$, to isolate the trend in the series of steel production figures given below.

Year	1975	1976	1977	1978	1979	1980	1981	1982	1983	1984
Production (millions of tonnes)	46	58	36	42	52	40	60	48	41	52

Sketch a graph of the production figures together with the smoothed values.

5. By using exponential smoothing with $\alpha = 0.25$ smooth out the weekly turnover figures for a company tabulated below.

Week	1	2	3	4	5	6	7	8	9	10	11	12	13	14
Turnover (×£100)	26	32	29	22	28	33	37	29	36	44	39	37	42	50

Illustrate the weekly turnover together with the smoothed values. Comment on whether these smoothed values give a reliable indication of the trend.

6.

Year	Domestic electricity consumption (million kilowatt hours)					
	Jan–Feb	Mar–Apr	May–June	July–Aug	Sept–Oct	Nov–Dec
1981	246	224	196	188	198	228
1982	248	219	192	189	201	232
1983	252	222	195	187	197	226

Use exponential smoothing with $\alpha = 0.2$ to isolate the trend. Hence, by using a multiplicative model, estimate the domestic electricity consumption during the first six months of 1984. You can assume that the previous two-monthly average consumption is 212 million kilowatt hours. Comment on the suitability of this forecasting method.

PAST QUESTIONS 6

1. The manager of a cycle and small motor-cycle business, examining his sales of mopeds over the past three years, finds the data to be as follows.

	No. of machines sold			
	1st quarter	2nd quarter	3rd quarter	4th quarter
1972	–	8	9	6
1973	10	15	18	8
1974	9	20	25	15
1975	13	22	–	–

(a) Use the data to make a reasonable estimate of the number of machines which should be ordered for the first half of 1977.

(b) What factors do you think are likely to affect the trend and seasonal variations in sales of mopeds? [IIM]

2. The booming tourist trade over the last few years has surpassed the expectations of a certain coach operator. It forces him into an urgent analysis of his operating policy over the next two years, for which he needs to make some short-term forecasts. His fleet's past performance is as follows (passenger miles in millions).

Quarter	1	2	3	4
1976	2.1	2.9	3.5	1.7
1977	2.7	4.3	4.5	2.1
1978	3.6	4.0	4.9	3.1

(a) Estimate the demand for all quarters of 1979 and the average yearly demand for 1980.

(b) What does the term 'seasonally adjusted' mean? [CIT]

3. The following data show the average monthly production of a resource in millions of kilograms for the years 1965–75.

1965	1966	1967	1968	1969	1970	1971	1972	1973	1974	1975
50.0	36.5	43.0	44.5	38.9	38.1	32.6	38.7	41.7	41.1	33.8

(a) Construct 3-year and 5-year moving averages for the time series and plot them on a graph together with the original data.

(b) Use this example to illustrate the potential uses of moving average methods in forecasting. [IIM]

4. The following data represent the number of 'vehicle days' (in thousands) per quarter which a large car hire company was able to sell during recent years.

Year	Quarter 1	Quarter 2	Quarter 3	Quarter 4
1977	4.2	5.8	7.0	3.4
1978	5.6	8.8	9.2	4.4
1979	7.2	7.6	9.4	5.8

(a) Produce forecasts for all quarters of 1980, using the method of moving averages.

(b) Instead of using moving averages, what other methods can be used for the determination of a trend line? Describe their advantages and disadvantages.

(c) The company accountant prefers to have his data in terms of 'revenue per quarter' rather than 'vehicle days'. What are the strengths and weaknesses of each form of presentation?

(d) What significant trends for the future can you see, and what would be the main features of a business forecast? What would be the best way of dealing with the seasonal fluctuations? [CIT]

5.

Quarter	× £1000				
	1975	1976	1977	1978	1979
Jan–Mar	30	38	39	43	46
Apr–June	11	13	18	18	19
July–Sept	20	18	21	23	–
Oct–Dec	42	47	50	54	–

A Manufacturer's Sales of Overcoats

Calculate the values for:

(a) the trend of the sales of overcoats,

(b) the regular seasonal movement of the sales using the method of moving averages. [IM]

6. Calculate the five-month moving average for the following sales information. Show both the original series and the calculated moving averages on the same graph. What purpose is served by such calculations and such a graph?

	Sales (× £1000)	
	1978	1979
January	24	46
February	39	38
March	41	30
April	31	37
May	20	49
June	20	36
July	42	34
August	47	32
September	36	40
October	25	50
November	32	42
December	44	37

[IM]

7. The following data show the yearly use by a clothing manufacturer of a resource for the years 1969 to 1979. The units are millions of kilograms.

1969	1970	1971	1972	1973	1974	1975	1976	1977	1978	1979
62.0	55.0	47.8	52.6	49.3	40.5	45.3	42.8	37.4	39.8	40.3

(a) Show how a three-year moving average could have been used to provide forecasts for the years 1972 to 1979, and determine the forecast errors using such a scheme of forecasting.

(b) Describe how an alternative forecasting scheme based on expo smoothing could have been used and discuss the relative merits above two forecasting methods.

8. The table shows the average household quarterly expenditure on all fuels for the years 1975-79. Identify the trend by the method of moving averages and by examining the fluctuations of the series about this trend, estimate the quarterly seasonal variations.

	1975	1976	1977	1978	1979
1st quarter	75	89	104	116	138
2nd quarter	42	51	59	66	79
3rd quarter	35	42	49	55	67
4th quarter	63	75	88	98	117

[IIM]

9. A company finds that the sales of its products are affected by the time of year, which tends to obscure any trend. The following sales data are available, showing sales in successive periods of four months.

Period	1976	1977	1978	1979
Jan–Apr	84	92	94	100
May–Aug	121	140	148	161
Sept–Dec	67	81	83	95

Required:

(a) Using the method of moving averages, estimate the trend in these data and calculate the 'seasonal' effects.

(b) Use your trend found in part (a) to estimate the mean monthly increase in sales that the company has achieved over the period.

(c) How much more would the company expect to sell in May-Aug, than in Sept-Dec in a situation where there is no trend? [ACA]

10. The following table shows the value of home and export orders for a manufacturer for the three years 1978-80. Each figure is expressed as an index based on 1977 = 100.

	Home orders			Export orders		
	1978	1979	1980	1978	1979	1980
1st quarter	97	104	106	98	107	117
2nd quarter	115	112	120	109	116	133
3rd quarter	109	111	117	105	111	129
4th quarter	93	99	105	95	105	115

(a) Calculate and tabulate the four quarterly moving averages for each set of data.

(b) Plot the observed indices on the same graph and superimpose both sets of moving averages.

(c) Comment briefly on any important features revealed by your graph.

[IIM]

11. Your company intends, as an economy measure, to reduce the number of general labourers from which manning to cover absenteeism on the production line is drawn. The production management are concerned about an adequate level of cover and to help decide on this, data on production line absenteeism has been collected. Analyse this data, comment on the absenteeism trends and seasonal variations and hence discuss any difficulties you might foresee in establishing the required level of cover economically.

Year	Quarter	Percentage absenteeism			
		1	2	3	4
1981		6.49	5.94	11.99	8.91
1980		5.72	5.06	10.78	7.81
1979		4.95	4.29	9.68	6.93
1978		4.29	3.74	8.69	6.27

[IPM]

12. A tour operator finds that his business is subject to seasonal variations. Expressed in terms of earned revenue (× £1000) per quarter, the picture looks as follows.

Year	Quarter 1	Quarter 2	Quarter 3	Quarter 4
1979	210	290	350	170
1980	280	440	460	220
1981	360	380	470	290

(a) Using the method of moving averages, produce estimates for all quarters of 1982.

(b) Are there any other methods which could have been used here?

(c) What are the implications for the tour operator both as regards the seasonal fluctuations and the future trend? [CIT]

13. The sales (× £1000) of golf equipment by a large department store is shown for each period of three months as follows.

Quarter	1978	1979	1980	1981
First	–	8	20	40
Second	–	30	50	62
Third	–	60	80	92
Fourth	24	20	40	–

(a) Using an additive model, find the centred moving average trend.

(b) Find the average seasonal variation for each quarter.

(c) Predict sales for the last quarter of 1981 and the first quarter of 1982, stating any assumptions. [ICMA]

14. (a) Plot a graph of the raw data tabulated opposite.

(b) Calculate appropriate moving averages, plot the trend, and from the trend estimate the turnover for each quarter of 1982.

(c) Explain the risks of using the figures calculated above as predictors for 1982.

% Turnover of the Labour Force

Year	Quarters			
	I	II	III	IV
1978	11	23	16	8
1979	14	29	16	9
1980	18	34	17	9
1981	19	42	23	12

[IMS]

15. Set out below are the premium figures for holiday insurance policies issued by an insurance company over the last five years on a quarter by quarter basis. The figures are given in £000's.

Year	1st quarter	2nd quarter	3rd quarter	4th quarter
1975	65	21	120	34
1976	71	21	132	36
1977	83	39	136	42
1978	89	44	148	59
1979	107	60	169	74

(a) Describe the components that you might expect to find in a time series such as this.

(b) Using any method that you consider appropriate, calculate figures for each quarter of 1980.

[CII]

16. (a) Describe briefly the methods that could be used to find the secular trend in time series data.

(b) Find the average seasonal deviations *and* indices for the data given below. The data are sales figures in £ with O being the original sales figures and T the corresponding trend figures.

Year	1st quarter		2nd quarter		3rd quarter		4th quarter	
	O	T	O	T	O	T	O	T
1978	207	231	200	249	188	269	576	290
1979	279	311	263	331	246	351	735	371
1980	350	390	329	410	302	430	894	449

Which is the better measure of seasonal variation for the above, the deviations or the indices and why?

[RICS]

7 INDEX NUMBERS

7.1 PRICE RELATIVES

Let p_i denote the current price of an item and p_0 the base price. The following formulae can be evaluated to indicate changes in prices.

(a) Simple price relative $= \dfrac{p_i}{p_0} \times 100$

(b) Average of price relatives $= \dfrac{\Sigma(p_i/p_0)}{n} \times 100$

(c) Weighted average of price relatives $= \dfrac{\Sigma w(p_i/p_0)}{\Sigma w} \times 100$

where w = weighting per item.

EXAMPLE 7.1.1 Highlight changes in the price of steel between 1980 and 1983 by finding indices based on 1980 prices.

Year	1980	1981	1982	1983
Average price of steel (£ per tonne)	300	315	326	336

SOLUTION We can use a simple price relative, i.e. $(p_i/p_0) \times 100$. Now, the base year is 1980. Therefore, price index for 1980 = <u>100</u>. The price indices for 1981–83 are calculated below.

1981: $\dfrac{315}{300} \times 100 = \underline{\underline{105}}$

1982: $\dfrac{326}{300} \times 100 = \underline{\underline{108.67}}$

1983: $\dfrac{336}{300} \times 100 = \underline{\underline{112}}$

Thus, the price of steel has increased by 12% between 1980 and 1983.

INDEX NUMBERS 137

EXAMPLE 7.1.2 Find the wages indices for 1981 and 1982 based on 1980 given the following information.

Year	1980	1981	1982
Average weekly wage (£)	112	122	130

SOLUTION Base date 1980. Therefore, wages index in 1980 = $\underline{100}$. The wages indices for 1981–82 are given below.

$$1981: \quad \frac{122}{112} \times 100 = \underline{108.93}$$

$$1982: \quad \frac{130}{112} \times 100 = \underline{116.07}$$

EXAMPLE 7.1.3 Find (a) fixed-based indices, and (b) chain-based indices for the value of exports achieved by a company during 1978–82.

Year	1978	1979	1980	1981	1982
Total exports (× £1000)	400	520	510	640	780

SOLUTION (a) *Fixed-based indices* (base date = 1978).

$$1978: \quad \underline{100}$$

$$1979: \quad \frac{520}{400} \times 100 = \underline{130}$$

$$1980: \quad \frac{510}{400} \times 100 = \underline{127.5}$$

$$1981: \quad \frac{640}{400} \times 100 = \underline{160}$$

$$1982: \quad \frac{780}{400} \times 100 = \underline{195}$$

(b) *Chain-based indices* (base date = previous period).

$$1978: \quad \underline{100}$$

$$1979: \quad \frac{520}{400} \times 100 = \underline{130}$$

$$1980: \quad \frac{510}{520} \times 100 = \underline{98.08}$$

$$1981: \quad \frac{640}{510} \times 100 = \underline{125.49}$$

$$1982: \quad \frac{780}{640} \times 100 = \underline{121.88}$$

EXAMPLE 7.1.4 Find the combined index of prices in 1983 based on those in 1981 for the commodities listed below.

Commodity	Price (£)	
	1981	1983
A	40	50
B	20	22
C	70	84

SOLUTION The price relatives for each commodity are:

(a)
$$\text{Commodity A:} \quad \frac{50}{40} \times 100 = 125$$

(b)
$$\text{Commodity B:} \quad \frac{22}{20} \times 100 = 110$$

(c)
$$\text{Commodity C:} \quad \frac{84}{70} \times 100 = 120$$

The combined index of prices can be obtained by finding the average of price relatives, i.e.

$$\frac{125 + 110 + 120}{3} = \underline{118.33}$$

EXAMPLE 7.1.5 Calculate an index of crop prices in 1982 based on 1981 given the table of information below.

Crop	Weighting	Price per tonne (£)	
		1981	1982
Wheat	60	120	156
Barley	30	105	110
Oats	10	80	92

SOLUTION The price indices (relatives) for each crop are calculated below.

$$\text{Wheat:} \quad \frac{156}{120} \times 100 = 130$$

$$\text{Barley:} \quad \frac{110}{105} \times 100 = 104.76$$

$$\text{Oats:} \quad \frac{92}{80} \times 100 = 115$$

INDEX NUMBERS   139

The weighted average of price relatives is:

$$\frac{\Sigma w(p_i/p_0) \times 100}{\Sigma w} = \frac{60 \times 130 + 30 \times 104.76 + 10 \times 115}{60 + 30 + 10}$$

$$= \frac{12\,092.8}{100} = \underline{120.93}$$

EXAMPLE 7.1.6 Evaluate a combined index of wage rates in 1983 based on 1980 figures from the following data.

Employee grade	Wage rates (£ per hour)		Estimated current annual expenditure (× £1000)
	1980	1983	
A	4.00	4.90	390
B	3.40	4.40	540
C	2.90	4.05	720

SOLUTION [*Note*: The weighting for each item is proportional to the total expenditure for that item.]

The calculations involved in finding the weighted average of price relatives are tabulated below.

Grade	p_0	p_i	w	$\dfrac{p_i}{p_0}$	$w\dfrac{p_i}{p_0}$
A	4.00	4.90	390	1.225	477.75
B	3.40	4.40	540	1.2941	698.814
C	2.90	4.05	720	1.3966	1005.552
			$\Sigma w = 1650$		$\Sigma w \dfrac{p_i}{p_0} = 2182.116$

$$\therefore \quad \text{Combined index} = \frac{\Sigma w(p_i/p_0)}{\Sigma w} \times 100$$

$$= \frac{2182.116}{1650} \times 100 = \underline{132.25}$$

7.2 PRICE AGGREGATES

It is usual for price indices to be calculated using one of the following aggregative formulae.

(a) Simple aggregate $= \dfrac{\Sigma p_i}{\Sigma p_0} \times 100$

(b) Laspeyre index $= \dfrac{\Sigma p_i q_0}{\Sigma p_0 q_0} \times 100 \qquad q_0 = $ base quantity

140 NOTES AND PROBLEMS IN STATISTICS

(c) Paasche index = $\dfrac{\Sigma p_i q_i}{\Sigma p_0 q_i} \times 100$ q_i = current quantity

(d) Typical year index = $\dfrac{\Sigma p_i q_t}{\Sigma p_0 q_t} \times 100$ q_t = 'typical' quantity

(e) Marshall–Edgeworth index = $\dfrac{\Sigma p_i (q_0 + q_i)}{\Sigma p_0 (q_0 + q_i)} \times 100$

(f) Fisher's ideal index = $\sqrt{\left(\dfrac{\Sigma p_i q_0}{\Sigma p_0 q_0}\right)\left(\dfrac{\Sigma p_i q_i}{\Sigma p_0 q_i}\right)} \times 100$

EXAMPLE 7.2.1 Find a combined index of share prices in 1983 based on 1980 values shown in the following table.

Company	Share prices (£)	
	1st January 1980	1st January 1983
X	0.80	0.90
Y	1.80	1.75
Z	2.40	2.85

SOLUTION

Simple aggregate index = $\dfrac{\Sigma p_i}{\Sigma p_0} \times 100$

$= \dfrac{0.90 + 1.75 + 2.85}{0.80 + 1.80 + 2.40} \times 100$

$= \dfrac{5.5}{5} \times 100$

$= \underline{110}$

EXAMPLE 7.2.2 Calculate the Laspeyre and Paasche indices of share prices in 1983 based on 1980 given the data tabulated.

Company	Share prices (£)		No. of shares sold	
	1st Jan 1980	1st Jan 1983	1st Jan 1980	1st Jan 1983
X	0.80	0.90	5000	10 000
Y	1.80	1.75	20 000	15 000
Z	2.40	2.85	40 000	60 000

SOLUTION

Laspeyre index = $\dfrac{\Sigma p_i q_0}{\Sigma p_0 q_0} \times 100$

$= \dfrac{0.90 \times 5000 + 1.75 \times 20\,000 + 2.85 \times 40\,000}{0.80 \times 5000 + 1.80 \times 20\,000 + 2.40 \times 40\,000} \times 100$

$= \dfrac{153\,500}{136\,000} \times 100$

$= \underline{112.87}$

$$\text{Paasche index} = \frac{\Sigma p_i q_i}{\Sigma p_0 q_i} \times 100$$

$$= \frac{0.90 \times 10\,000 + 1.75 \times 15\,000 + 2.85 \times 60\,000}{0.80 \times 10\,000 + 1.80 \times 15\,000 + 2.40 \times 60\,000} \times 100$$

$$= \frac{206\,250}{179\,000} \times 100$$

$$= \underline{115.22}$$

EXAMPLE 7.2.3 Calculate the Laspeyre and Paasche price indices for 1982 given the data shown below (base date = 1981).

Item	1981 Price (£)	1981 Quantity purchased	1982 Price (£)	1982 Quantity purchased
I	1.50	20	1.60	20
II	0.80	40	0.85	60
III	0.40	60	0.55	50
IV	2.00	100	2.25	90

SOLUTION The required totals are calculated in the following table.

p_0	q_0	p_i	q_i	$p_0 q_0$	$p_i q_0$	$p_0 q_i$	$p_i q_i$
1.5	20	1.6	20	30	32	30	32
0.8	40	0.85	60	32	34	48	51
0.4	60	0.55	50	24	33	20	27.5
2.0	100	2.25	90	200	225	180	202.5
				286	324	278	313

$$\text{Laspeyre price index} = \frac{\Sigma p_i q_0}{\Sigma p_0 q_0} \times 100 = \frac{324}{286} \times 100 = \underline{113.29}$$

$$\text{Paasche price index} = \frac{\Sigma p_i q_i}{\Sigma p_0 q_i} \times 100 = \frac{313}{278} \times 100 = \underline{112.59}$$

[*Note*: In practice, usually only one of these indices would be calculated.]

EXAMPLE 7.2.4 The table below gives the retail car prices together with the number of models sold during the first week of August in 1980 and 1982.

Model	1980 Price	1980 No. sold (thousands)	1982 Price	1982 No. sold (thousands)
A	£3600	12	£4000	18
B	£4500	10	£4800	20
C	£6000	8	£8000	6

Illustrate the overall change in car prices between 1980 and 1982 by calculating:

(a) the Marshall–Edgeworth index, and

(b) Fisher's ideal index.

SOLUTION (a)

p_0	q_0	p_i	q_i	q_0+q_i	$p_0(q_0+q_i)$	$p_i(q_0+q_i)$
3600	12	4000	18	30	108 000	120 000
4500	10	4800	20	30	135 000	144 000
6000	8	8000	6	14	84 000	112 000
					327 000	376 000

$$\text{Marshall–Edgeworth index} = \frac{\Sigma p_i(q_0+q_i)}{\Sigma p_0(q_0+q_i)} \times 100$$

$$= \frac{376\,000}{327\,000} \times 100$$

$$= \underline{114.98}$$

(b)

p_0	q_0	p_i	q_i	p_0q_0	p_iq_0	p_0q_i	p_iq_i
3600	12	4000	18	43 200	48 000	64 800	72 000
4500	10	4800	20	45 000	48 000	90 000	96 000
6000	8	8000	6	48 000	64 000	36 000	48 000
				136 200	160 000	190 800	216 000

$$\text{Fisher's ideal index} = \sqrt{\left(\frac{\Sigma p_iq_0}{\Sigma p_0q_0}\right)\left(\frac{\Sigma p_iq_i}{\Sigma p_0q_i}\right)} \times 100$$

$$= \sqrt{\left(\frac{160\,000}{136\,200}\right)\left(\frac{216\,000}{190\,800}\right)} \times 100$$

$$= \sqrt{(1.1747)(1.1321)} \times 100$$

$$= 1.1532 \times 100$$

$$= \underline{115.32}$$

[*Note*: In general, such indices would only be calculated if the simpler indices — e.g. Laspeyre and Paasche — were found to be unsuitable.]

EXAMPLE 7.2.5 Calculate indices of the tuition fees in a college of further education for 1982 and 1983 based on 1981.

Type of course	Average tuition fees per student (£)		
	1981	1982	1983
Full-time	1400	1900	2200
Part-time	110	140	140
Short	30	50	60

INDEX NUMBERS   143

It can be assumed that in a typical year for every 10 full-time students, there are 16 part-time and 3 short-course students attending this college.

SOLUTION The calculations are tabulated below.

p_0	p_1	p_2	q_t	$p_0 q_t$	$p_1 q_t$	$p_2 q_t$
1400	1900	2200	10	14 000	19 000	22 000
110	140	140	16	1760	2240	2240
30	50	60	3	90	150	180
				15 850	21 390	24 420

The typical year indices for 1982 and 1983 based on 1981 are as follows.

$$1982: \frac{\Sigma p_1 q_t}{\Sigma p_0 q_t} \times 100 = \frac{21\,390}{15\,850} \times 100$$

$$= \underline{134.95}$$

$$1983: \frac{\Sigma p_2 q_t}{\Sigma p_0 q_t} \times 100 = \frac{24\,420}{15\,850} \times 100$$

$$= \underline{154.07}$$

7.3 QUANTITY/VOLUME INDICES

The formulae for quantity indices can be obtained by interchanging the values of p and q in the price index formulae, e.g.

(a) Simple aggregate quantity index $= \dfrac{\Sigma q_i}{\Sigma q_0} \times 100$

(b) Laspeyre quantity index $= \dfrac{\Sigma q_i p_0}{\Sigma q_0 p_0} \times 100$

(c) Paasche quantity index $= \dfrac{\Sigma q_i p_i}{\Sigma q_0 p_i} \times 100$

EXAMPLE 7.3.1 Calculate indices based on 1980 for the volume of wheat exported given the table below.

Year	1980	1981	1982	1983
Total export (1000 tonnes)	56	64	66	62

SOLUTION The volume (or quantity) indices for 1981–83 are calculated as follows.

$$1981: \frac{64}{56} \times 100 = \underline{114.29}$$

$$1982: \frac{66}{56} \times 100 = \underline{117.86}$$

$$1983: \frac{62}{56} \times 100 = \underline{110.71}$$

EXAMPLE 7.3.2 Calculate a combined index of steel production for 1982 based on 1980 in the three factories tabulated below.

Factory	Production (1000 tonnes)	
	1980	1982
A	80	82
B	120	110
C	40	50

SOLUTION

$$\text{Quantity index} = \frac{\Sigma q_i}{\Sigma q_0} \times 100$$

$$= \frac{82 + 110 + 50}{80 + 120 + 40} \times 100$$

$$= \frac{242}{240} \times 100$$

$$= \underline{100.83}$$

EXAMPLE 7.3.3 Calculate a quantity index for February 1982 based on the figures in January 1982 shown below.

Commodity	Price (£)	Quantity purchased	
		Jan 1982	Feb 1982
A	3.50	1200	1400
B	10.00	700	650
C	2.40	1500	1400
D	6.20	300	420

SOLUTION The required summations are calculated in the following table.

q_0	q_i	p	$q_0 p$	$q_i p$
1200	1400	3.50	4200	4900
700	650	10.00	7000	6500
1500	1400	2.40	3600	3360
300	420	6.20	1860	2604
			16 660	17 364

$$\text{Quantity index} = \frac{\Sigma q_i p}{\Sigma q_0 p} \times 100$$

$$= \frac{17\ 364}{16\ 660} \times 100$$

$$= \underline{104.23}$$

EXAMPLE 7.3.4 Find (a) a Laspeyre index, and (b) a Paasche index of the quantity produced in 1983 based on the figures in 1978.

Product	No. of units produced		Cost per unit (£)	
	1978	1983	1978	1983
X	3000	2600	7.00	16.00
Y	5000	6000	5.00	9.50
Z	10 000	14 000	0.80	1.20

SOLUTION

q_0	q_i	p_0	p_i	$q_0 p_0$	$q_i p_0$	$q_0 p_i$	$q_i p_i$
3000	2600	7	16	21 000	18 200	48 000	41 600
5000	6000	5	9.5	25 000	30 000	47 500	57 000
10 000	14 000	0.8	1.2	8000	11 200	12 000	16 800
				54 000	59 400	107 500	115 400

(a) Laspeyre quantity index $= \dfrac{\Sigma q_i p_0}{\Sigma q_0 p_0} \times 100$

$= \dfrac{59\,400}{54\,000} \times 100$

$= \underline{110}$

(b) Paasche quantity index $= \dfrac{\Sigma q_i p_i}{\Sigma q_0 p_i} \times 100$

$= \dfrac{115\,400}{107\,500} \times 100$

$= \underline{107.35}$

EXERCISES ON SECTIONS 7.1 TO 7.3

1. Calculate the salary indices for 1980–83 based on 1979 given the following information.

Year	1979	1980	1981	1982	1983
Average monthly salary (£)	450	524	592	650	705

2. Calculate (a) fixed-based price indices, and (b) chain-based price indices, for the data given below.

(a)

Year	1978	1979	1980	1981	1982
Average house prices (£1000)	16	23	28	27	32

(b)

Month	January	February	March	April	May	June
Price of gold ($)	400	420	445	430	425	440

3. Find a combined index of prices in 1983 based on 1982 for each group of commodities given below.

(a)

Commodity	Price (£) 1982	Price (£) 1983
A	34	38
B	10	10
C	26	29
D	3	4

(b)

Commodity	Weighting	Price (£) 1982	Price (£) 1983
X	14	50	54
Y	20	8	10
Z	6	26	25

4. Evaluate (i) a Laspeyre price index, and (ii) a Paasche price index for 1982 based on 1980 for each group of items tabulated below.

(a)

Item	1980 Price (£)	1980 Quantity	1982 Price (£)	1982 Quantity
A	40	6	46	8
B	20	12	22	15
C	8	20	12	18

(b)

Item	Price (£) 1980	Price (£) 1982	Quantity 1980	Quantity 1982
X	0.80	0.98	100	100
Y	1.20	1.50	80	70
Z	3.00	2.80	40	90

5. Calculate the (a) Laspeyre index, (b) Paasche index, and (c) Marshall–Edgeworth index of wage rates in 1983 based on 1982 from the following data.

Grade of employee	Hourly rate (£)		No. of employees	
	1982	1983	1982	1983
I	2.20	2.50	120	80
II	2.70	3.00	140	150
III	3.10	3.30	80	100

6. Find a suitable index of the amount of overtime earned during 1983 based on 1980 figures in the three factories given below.

Factory	Average overtime earned per week (£)		No. of employees	
	1980	1983	1980	1983
A	30	20	600	500
B	40	45	800	850
C	25	30	1000	1000

7. Calculate indices of the volume of exports in 1981 and 1982 based on 1980 from the following information.

Company	Volume of exports (no. of units)		
	1980	1981	1982
I	300	360	400
II	500	400	460
III	600	620	640
IV	400	380	360

8. Calculate the Laspeyre and Paasche indices of quantities in 1984 based on 1983 values shown in the following table.

Commodity	No. of items purchased		Price per item (£)	
	1983	1984	1983	1984
X	600	625	320	330
Y	2000	2300	110	140
Z	1400	1545	160	150

9. Find the Laspeyre and Paasche indices of both prices and quantities in 1982 based on 1981 values shown in the following table.

Commodity	Total imports			
	Quantity		Price (£ millions per unit)	
	1981	1982	1981	1982
Oil (million barrels)	70	64	35	32
Steel (100 000 tonnes)	6	10	28	29
Timber (10 000 tonnes)	90	92	1.4	1.7

PAST QUESTIONS 7

1. A factory produces togs, clogs and pegs, each of these three products having a different work content. The proportions of these products vary from month to month and the factory requires an index for assessing productivity changes. Each tog, clog and peg produced is to be weighted according to its work content, these weights being 6, 8 and 5 respectively. Also, because some months contain more working days than others the index should offset the effect of this.

   Data for the months of May, June and July are as follows.

	May	June	July
No. of working days	23	22	16 (due to factory closure for two weeks)
Output (thousands)			
togs	19	16	10
clogs	12	20	15
pegs	22	15	10

   It is intended that May should be the base month for comparison, with a productivity index of 100.

   Required:

   (a) Design a simple productivity index, calculate its value for June and July, and comment briefly on the results.

   (b) Now, due to a change in the type of peg produced, a new weight is required. Production data are shown below for two days when productivity was judged to be about equal.

Output	Day 1	Day 2
togs	921	811
clogs	800	747
new pegs	1042	1206

   Use these data to estimate a suitable weight for the new pegs, to 1 decimal place, assuming that the weightings of 6 for togs and 8 for clogs are as before. [ACA]

2. From the data below calculate a base weighted and a current weighted all items index for 1979 with 1956 = 100. Comment upon your results. If the average weekly wage for an adult male manual worker was £12 in 1956 and £80 in 1979 estimate the change in the standard of living between the two years.

	1956		1979	
	Prices	Weights	Prices	Weights
Food	12.50	350	18.00	230
Housing	7.00	90	14.00	120
Fuel and light	2.50	55	4.50	60
Clothing	6.00	105	7.50	60

   [RVA]

3. What is an index number?

   Explain the problems that face one in choosing the items for inclusion in a price index.

   Calculate a Laspeyre price index for December 1978 from the following data, using January 1974 as a base.

Commodity	January 1974		December 1978	
	Price (£)	Quantity	Price (£)	Quantity
A	1.62	34	1.95	32
B	3.40	20	4.75	25
C	0.95	12	2.55	11
D	5.80	4	9.45	2

   [IM]

4. The table below shows the price of each of five materials purchased by a local authority for house maintenance and the proportionate demand that the purchase of each of these made to the total cost of house maintenance.

		Material				
		A	B	C	D	E
1970	Price per unit £	1	2	3	5	8
	% of purchases	15	10	30	25	20
1980	Price per unit £	2	3	5	8	12
	% of purchases	10	15	40	20	15

   Calculate an index number to measure the overall change in the prices of these materials between 1970 and 1980. Discuss your result and in particular the assumptions that have to be made in its interpretation. [RVA]

5. (a) Define what is meant by a 'fixed base index number' and a 'chain-based index number' and explain the different ways in which these alternatives have to be interpreted.

   (b) For the following data calculate:
   (i) a Laspeyre *price* index for 1980,
   (ii) a Paasche *quantity* index for 1980,
   in each case using 1978 as the base year.

Commodity	1978		1980	
	Average price (£)	Quantity	Average price (£)	Quantity
A	3.65	156	3.75	194
B	7.82	274	9.20	305
C	1.40	115	1.80	187
D	2.95	432	4.54	378
E	14.84	89	20.36	126

   [IM]

6. Tixif Limited sells three types of chain-saws to the public. Company records show the prices and quantities sold of each type were as follows.

Type	1979		1980	
	Price	No. sold	Price	No. sold
X	30	22	40	30
Y	50	31	60	40
Z	120	8	99	12

(a) Calculate an unweighted aggregate price index for 1980, with 1979 = 100.

(b) Calculate a weighted Laspeyre aggregate price index for 1980, with 1979 = 100.

(c) The owner of Tixif Limited wants to know whether or not the average price paid for his chain-saws has increased or decreased, 1979–80, and by how much. What would you tell him? [ICMA]

7. (a) Discuss the problems of identifying suitable weights for an index number, giving appropriate examples from published indices.

(b) From the following information of the Transport and Vehicles sub-group of the Retail Price Index calculate a weighted price-relative index for 1980 based on 1974 prices.

	Weight	1974 Price (£)	1980 Price (£)
Purchase of motor vehicles	56	1731	4623
Maintenance of motor vehicles	15	50	158
Petrol and oil	43	114	372
Motor licences	7	36	85
Motor insurance	10	68	184
Rail transport fares	8	20	69
Road transport fares	12	8	26

[IM]

8.

Type of goods purchased	Weekly expenditure		Relative quantity purchased	
	1960 £	1980 £	1960 %	1980 %
Food and drink	8.40	15.68	35	23
Housing and fuel	3.00	5.40	25	30
Goods and clothing	1.20	2.40	20	15
Transport and services	4.00	8.40	20	32

Calculate the current weighted and base weighted index number for the overall prices in 1980 on a base of 1960.

Discuss generally the value of such indices to the rating or valuation officer and, in particular, the effect of different weighting systems upon the results.

[RVA]

9. Prodco Limited is considering constructing a weighted index of prices for the components P, Q, R and S of its stock. The weightings of the components together with the prices for the years 1976-80 are shown in the following table.

Component	Weighting of component	Prices (£)				
		1976	1977	1978	1979	1980
P	10	6.00	7.50	9.25	11.00	13.50
Q	12	3.00	4.50	6.00	8.00	10.50
R	17	2.50	4.00	5.80	7.00	9.20
S	9	4.75	6.00	7.25	8.75	10.00

Required:

(a) Calculate a series of weighted price index numbers (correct to one place of decimals) for 1977 to 1980 based on the 1976 figures.

(b) Compare and contrast the calculation of a set of index numbers if the Laspeyre and Paasche Price Index Numbers are calculated.

(c) Explain what practical problems are likely to be encountered in establishing an index number to measure changes in consumer prices. [ACA]

10. (a) Explain what is meant by an index number. Discuss the different ways it may be calculated and its use in personnel management decision-making.

(b) Your organisation employs five grades of blue-collar workers. Numbers and rates for three years are shown in the table below.

Grade	1979		1980		1981	
	Rate/hr	No. of employees	Rate/hr	No. of employees	Rate/hr	No. of employees
1	2.00	30	2.20	34	2.44	37
2	1.86	16	2.04	14	2.26	11
3	1.68	15	1.85	14	2.05	13
4	1.32	45	1.47	44	1.63	46
5	1.23	20	1.37	20	1.52	18

Taking 1979 as base, calculate an index number of the average wage for 1980 and 1981. Explain why you have chosen the formula you have used to calculate the index. [IPM]

11. (a) What is an index number? Explain the problems that face one in choosing the items for inclusion in the compilation of a price index.

(b) Calculate the December 1981 Paasche price index for the group of commodities given taking January 1978 as a base and using such data as is required from the following table.

Commodity	January 1978		December 1981	
	Price (£)	Quantity	Price (£)	Quantity
A	1.83	18	1.97	22
B	2.42	35	2.84	38
C	3.85	27	2.95	31
D	1.67	15	2.03	19
E	0.75	58	0.95	63
F	2.46	31	2.78	20

[IM]

12. (a) Briefly explain two commercial, industrial or business uses of index numbers.

(b) A cost accountant has derived the following information about weekly wage rates (W) and the number of people employed (E) in the factories of a large chemical company.

Technical group of employees	July 1979		July 1980		July 1981	
	W	E	W	E	W	E
Q	60	5	70	4	80	4
R	60	2	65	3	70	3
S	70	2	85	2	90	1
T	90	1	110	1	120	2

Basic weekly rates (£'s) and number of employees (100's)

(a) Calculate a Laspeyre (base weights) all-items index number for the July 1980 basic weekly wage rates, with July 1979 = 100.

(b) Calculate a Paasche (current weights) all-items index number for the July 1981 basic weekly wage rates, with July 1979 = 100.

(c) Briefly compare your index numbers for the company with the official government figures for the Chemical and Allied Industries which are given below.

	Yearly annual averages		
	1979	1980	1981
Weekly wage rates (July 1976 = 100)	156.3	187.4	203.4

[ICMA]

13.

Expenditure	1974		1982	
	Weights	Prices	Weights	Prices
Rent	27	79	32	86
Mortgage interest	40	96	31	161
Rates and water	34	48	34	48
Repairs and maintenance	19	56	23	69

Calculate an all-items index, base 1974 = 100, first, base weighted and then current weighted, for 1982. Discuss the advantages of the two indices. [RVA]

14. (a) Discuss the factors to be considered when constructing an index number series.

(b) A supplier of building materials wishes to assess the overall change in both prices and quantities over the past three years for the following.

Item	Unit	Price (£)			Quantity		
		1978	1979	1980	1978	1979	1980
A	m^2	5	7	10	10	10	14
B	tonne	20	22	26	20	30	25
C	m^3	4	8	12	100	110	80
D	tonne	10	16	20	30	34	20

Construct either a Laspeyre or a Paasche type index series for both prices and quantities, stating which type you are using. Construct a value index series from the data. [RICS]

15. (a) What are the main factors to be considered when constructing an index number series?

(b) The following data gives the quantities and unit prices for four items over the last three years. Use them to calculate:
   (i) a Laspeyre index series for quantities,
   (ii) a Paasche index series for prices.

Item	Unit	Quantity			Price (£)		
		1976	1977	1978	1976	1977	1978
A	tonne	5	8	12	2.00	2.10	2.16
B	m^2	5	6	7	1.00	1.50	2.50
C	m^3	2	4	5	4.00	4.00	4.20
D	tonne	4	4	6	3.00	3.30	3.60

[RICS]

ANSWERS

SECTIONS 1.1 TO 1.5

2. (a) 61 days (b) 27 days.

3.

No. of minutes late	0–	10–	20–	30–	40–	50–
No. of employees at A	24	8	7	3	2	1
No. of employees at B	23	19	13	4	4	2

4.

Time lost (min)	0–	10–	20–	30–	40–	50–	60–	70–	80–
No. of days	3	2	12	23	11	4	2	0	3

SECTIONS 1.6 TO 1.8

1. (a) (i) 5.24 ✓ (ii) 5.2 ✓ (iii) 5.1 ✓
   (b) (i) 23.5 (ii) 21 (iii) 18.

2. 32.5 years, 26.5 years.

3. 310.2 units.

4. 13 faulty chips, 56%.

SECTIONS 1.9 TO 1.13

1. (a) (i) 10 (ii) 4.89 (iii) 6.60.
   (b) (i) 5.5 (ii) 4.32 (iii) 6.48.

2. £117.75, £12.04.

3. £15.50, £15.

4. 10.88 months, 12.03 months.

5. 1272.92, 43.48 units.

PAST QUESTIONS 1

1. (a) 5560 hours, 1455 hours (b) $4\frac{1}{2}\%$.

2. (a) 50 minutes (b) 68 minutes.

3. (a) (i) 580 (ii) 250 (b) 575 (c) 108.

4. (a) (i) 64 (ii) 34 (iii) 36 (iv) 16 (v) 52 (vi) 18 (vii) 17.2
   (viii) 19.63.

ANSWERS 155

6. (a)

Exports	50-59	60-69	70-79	80-89	90-99	100-109	110-119	120-129	130-139
Frequency	3	7	15	18	21	14	6	2	2

(b) 89.82, 89.84   (c) 17.12.

7. (a) A: 15.346, 2.710; B: 13.654, 3.589.

8. (a)

No. of tickets	2	3	4	5
Frequency	83	82	30	5

Denominations (p)	1	2	5	13
Frequency	62	138	198	159

(b) (i) 2.785   (ii) 3   (iii) 2   (iv) 35.5%.

9. (a)

Kilos	0-	10-	20-	30-	40-	50-	60-	70-	80-
A	16	12	10	5	4	1	1	1	0
B	17	18	6	3	1	3	1	0	1

(c) A: median = 18, IQR = 22;   B: median = 14, IQR = 16.

10. (a)

Rejects	1-5	6-10	11-15	16-20	21-25	26-30	31-35
Frequency	2	3	5	8	20	10	2

(b) (i) Mean = 20.9   (ii) standard deviation = 6.935.

12. Q_1 = £7000, Q_3 = £8800, median = £8200.

13. USA: 83, 42, 120, 39; Japan: 14, 8, 26, 9.

14. (a)

Price	105-	110-	115-	120-	125-	130-	135-	140-
Frequency	2	5	4	8	10	5	4	2

(b) 125.5 p, 119 p, 131 p, 6 p   (c) 124.5 p, 9.01 p.

15. (a)

Hours	20-	25-	30-	35-	40-	45-
Frequency	1	2	14	14	8	1

16.

Income (× £1000)	0-	200-	400-	600-	800-	1000-	1200-
Frequency	6	5	8	5	4	0	2

(a) £526 670   (b) £260 450.

17. (b) Mean: A = 25.64 mm, B = 23.80 mm; A costs £6.60, B costs £13.50.

SECTIONS 2.1 AND 2.2

1. (a) 0.9   (b) 0.3   (c) 0.07   (d) 0.27.
2. (a) 0.6   (b) 0.5   (c) 0.9.
3. (a) 0.81   (b) 0.18   (c) 0.01.
4. (a) $\frac{4}{9}$ (b) $\frac{4}{9}$
5. (a) 0.512 (b) 0.096.
6. (a) 0.49   (b) 0.495.

SECTIONS 2.3 TO 2.5

1. (a) 10   (b) 7   (c) 1   (d) 1.
2. (a) 0.168 07   (b) 0.308 70   (c) 0.836 92.
3. (a) 0.082 08   (b) 0.287 30   (c) 0.456 19.
4. (a) 0.531 44   (b) 0.354 29   (c) 0.114 27.
5. (a) 0.316 41   (b) 0.050 78, 1, 0.8660.
6. (a) 0.301 19   (b) 0.120 51.
7. (a) 0.548 81   (b) 0.121 90   (c) 0.017 35.

SECTION 2.6

1. (a) 0.158 66   (b) 0.579 26   (c) 0.725 75   (d) 0.3057
   (e) 0.6687
2. (a) (i) 0.0304   (ii) 0.158 66   (iii) 0.6284   (b) 19.32 to 50.68.
3. (a) 0.3286   (b) 30   (c) £1.70.
4. (a) 0.0478   (b) 10.62 ohms   (c) 20.
5. (a) 0.0161   (b) 8 weeks.

PAST QUESTIONS 2

1. (a)

Output	0.57	0.58	0.59	0.60	0.61	0.62	0.63	0.64
Frequency	1	2	5	10	9	4	3	2

   Mean = 0.6061, standard deviation = 0.0160.
   (b) 0.5747 to 0.6375 kW.
2. (a) 0.857 12   (b) 7.07 defectives.
3. (a) (i) 1. $\frac{1}{48}$ 2. $\frac{1}{4}$ 3. $\frac{35}{48}$ (ii) sum = 1
   (b) (ii) £23.30.

4. (a) (i) 0.590 49 (ii) 0.328 05 (iii) 0.000 46
   (b) (i) 0.5 (ii) 0.0038 (iii) 0.000 0002.

5. (a) 0.774 53 (b) 40.53 oz (c) £7.16 (d) 41.861 oz.

6. (a) 28.4% (b) 318 hours (c) 0.847.

7. (a) (i) 0.135 34 (ii) 0.323 32 (iii) 0.135 34
   (b) (i) 0.82% (ii) 5.48%.

8. (a) (i) 0.579 26 (ii) 0.3674 (b) 60 marks (c) 66.7%
   (d) 55.33 marks.

9. (b) (i) 0.9181 (ii) 0.7788 (iii) 680 components.

10. (a) (i) 0.496 (ii) 0.398 (b) (i) 0.367 88 (ii) 0.264 24.

11. (b) (i) 0.0567 (ii) 0.1230 (normal approximation to Poisson).

12. (a) (i) 46.82 hours (ii) 26.1% (b) (i) 0.008 56 (ii) No.

13. (a) 0.7745 (ii) 0.066 81 (iii) 0.1359 (iv) 0.158 66.

14. (a) 0.590 49 (b) 0.000 45 (c) 0.918 54 4.5, 0.6708.

15. 0.022 75, 0.006 21, 0.5328.

16. (b) (i) 0.036 (ii) 0.084 (iii) 0.302 (iv) 0.614.

SECTIONS 3.1 AND 3.2

1. 300, 5.

2. (a) 0.105 65 (b) 0.4680.

3. (a) 0.0045 (b) 0.0410 (c) 0.2357.

4. (a) (i) 0.1508 (ii) 0.1498 (b) £3710.22 to £4089.78.

5. $u = 2.108$, significant (2 tails).

6. $u = 2.556$, significant (1 tail).

7. $u = 1.905$, significant (1 tail).

SECTIONS 3.3 TO 3.5

1. (a) 2.920 (b) 2.015 (c) 1.697.

2. $t = 2.937$, $\nu = 29$, significant.

3. $t = 3.333$, $\nu = 64$, significant.

4. $t = 1.715$, $\nu = 7$, not significant.

5. $t = 0.667$, $\nu = 5$, not significant.

6. 24.099 8.

7. $t = 5.702$, $\nu = 30$, significant.

8. $t = 2.097$, $\nu = 88$, not significant.
9. $t = 2.150$, $\nu = 40$, significant.
10. $t = 2.117$, $\nu = 7$, not significant (5% level).
11. $t = 2.910$, $\nu = 8$, significant (1% level, 1 tail).

SECTION 3.6

1. (a) 0.5, 0.05 (b) 0.3, 0.059.
2. 0.339 to 0.461.
3. (a) 0.0155 (b) 0.1778.
4. 0.521 to 0.629.
5. $u = 1.456$, not significant.
6. $u = 1.225$, not significant.
7. $u = 1.664$, not significant.
8. $u = 1.874$, significant.

PAST QUESTIONS 3

1. (a) $t = 6.928$, $\nu = 48$, significant (b) $t = 1.876$, $\nu = 84$, not significant.
2. (a) $t = 2.976$, $\nu = 63$, significant (b) 0.271.
3. (a) $u = 3.078$, significant (b) $t = 8.888$, $\nu = 79$, significant.
4. (a) 55% (b) $u = 2.274$, significant (c) over 18% difference.
5. (b) (i) $t = 3.266$, $\nu = 8$, significant.
6. (a) (i) 0.590 49 (ii) 0.081 46 (b) $u = 1.043$, not significant.
7. (b) (i) 0.0210 (ii) $u = 2.619$, significant.
8. (b) $u = 1.81$, not significant.
9. (a) $u = 1.833$, significant (b) $u = 1.833$, not significant.
10. (b) $t = 4.596$, significant.
11. 41.95 to 43.69 years.
12. $t = 2.507$, $\nu = 49$, significant.
13. $t = 2.358$, $\nu = 22$, significant.
14. (a) $u = 1.705$, not significant, $u = 1.046$, not significant
    (b) $u = 2.412$, significant, $u = 1.479$, not significant.

SECTION 4.1

1. (a) 12.5916 (b) 24.9958 (c) 43.7730.

2. 0.206 989 to 14.8603.
3. (a) $\chi^2 = 2.258$, not significant (b) $\chi^2 = 4.087$, not significant
   (c) $\chi^2 = 9.064$, not significant.
4. $\chi^2 = 12.695$, $\nu = 3$, significant.
5. $\chi^2 = 2.644$, $\nu = 4$, not significant.
6. $\chi^2 = 0.774$, $\nu = 2$, not significant.

SECTIONS 4.2 AND 4.3

1. (a) 2   (b) 8   (c) 20.
2. (a) $\chi^2 = 3.056$, $\nu = 2$, not significant (b) $\chi^2 = 15.650$, $\nu = 4$, significant
   (c) $\chi^2 = 5.956$, $\nu = 1$, significant.
3. $\chi^2 = 5.385$, $\nu = 3$, not significant.
4. $\chi^2 = 7.164$, $\nu = 2$, not significant.
5. $\chi^2 = 5.406$, $\nu = 1$, significant.

PAST QUESTIONS 4

1. $\chi^2 = 3.75$, $\nu = 2$, not significant.
2. (a) $\chi^2 = 5.316$, $\nu = 4$, not significant (b) $\chi^2 = 13.504$, $\nu = 2$, significant.
3. (a) $\chi^2 = 97.09$, $\nu = 3$, significant.
4. (b) $\chi^2 = 3.901$, $\nu = 2$, not significant.
5. (a) $\chi^2 = 6.482$, $\nu = 4$, not significant.
6. (b) $\chi^2 = 35.30$, $\nu = 6$, significant.
7. $\chi^2 = 4.719$, $\nu = 6$, not significant.
8. (a) $\chi^2 = 3.683$, $\nu = 1$, not significant.
9. $\chi^2 = 5.483$, $\nu = 4$, not significant.
10. (b) $\chi^2 = 12.936$, $\nu = 4$, significant.
11. $\chi^2 = 51.37$, $\nu = 4$, significant.
12. $\chi^2 = 0.127$, $\nu = 2$, not significant.
13. $\chi^2 = 3$, $\nu = 4$, not significant.
14. $\chi^2 = 1.621$, $\nu = 2$, not significant.

SECTIONS 5.1 TO 5.3

1. (a) $r = 1$ (b) $r = -1$ (c) $r = 0$.
2. $r = 0.951$.

3. $r = 0.85$.

4. (a) 0.650   (b) 0.779.

SECTIONS 5.4 AND 5.5

1. (a) $r = 0.981$, $y = 0.9 + 0.85x$, $y = 11$
(b) $r = -0.984$, $y = 13.6 - 0.6x$, $y = 8$
(c) $r = -0.363$, ($t = 0.935$), not significant.

2. $r = 0.970$, $y = 3.6 + 3.667x$, £180 000.

3. $r = 0.928$, $y = 3.22 + 0.651x$ ($x = $ year $- 1976$), £7800.

4. $r = -0.873$ ($t = 5.063$, significant), $y = 185.2 - 38.687x$, 88 hours.

PAST QUESTIONS 5

1. (c) $r = +0.950$.

2. (a) $r = 0.974$ (b) $y = 0.565 + 0.620x$, 6×10^7 BTU.

3. (a) (ii) 26 minutes.

4. (b) $r = 0.857$.

5. (b) $r = 0.476$, not significant.

6. (a) (i) $y = 13.3 + 1.0727x$ ($x = $ year $- 1970$)   (ii) 14.4, 24.0
(b) £25.1 million.

7. (c) (ii) $r = -0.523$.

8. (b) $r = 0.4286$ (1, 2); $r = 0.7714$ (1, 3); $r = 0.7143$ (2, 3). (N.B.: $r = R_S$.)

9. (a) $r = -0.879$ (b) $y = 27.88 - 0.39x$.

10. (b) $y = 0.366 + 0.232x$ (c) £3810.

11. (b) $Y = 210 + 3X$ (d) £330.

12. $r = -0.858$, $t = 5.282$, $v = 10$, significant.

13. (b) $r = -0.871$ (c) $Y = 24.89 - 0.9X$, 24 rejects.

14. $v = 21.292$, $f = 26.126$, $r = 0.994$.

15. (b) $y = 58.11 - 2.12x$.

16. $r = 0.457$, not significant.

17. $r = -0.8$, significant.

18. $r = 0.795$, significant.

SECTIONS 6.1 TO 6.3

1. (a) 2.73, 2.8, 2.93, 3.17, 3.37, 3.5, 3.6, 3.77, 4.03, 4.27, 4.5
(b) 3.014, 3.143, 3.314, 3.457, 3.671, 3.886, 4.029.

2. £44 000.

3. 5.3, 6.4, 5.6.

4. (a) 34, 47, 34, 34   (b) 29, 24, 25, 33.

5. 421, 532, 540, 435 (£ millions).

SECTIONS 6.4 AND 6.5

1. (a) multiplicative   (b) additive   (c) multiplicative.

2. 740, 440, 720 (multiplicative).

3. 620, 1100, 1200, 750 (multiplicative).

4. (a) 46, 47.2, 46.08, 45.67, 46.30, 45.67, 47.10, 47.19, 46.57, 47.11
   (b) 46, 50.8, 44.88, 43.73, 47.04, 44.22, 50.53, 49.52, 46.11, 48.47.

5. 26, 27.5, 27.875, 26.406, 26.805, 28.354, 30.516, 30.137, 31.603, 34.702, 35.777, 36.083, 37.562, 40.672.

6. 656 million kWh.

PAST QUESTIONS 6

1. (a) 57 mopeds (multiplicative).

2. (a) 1979: 4.1, 5.0, 5.5, 3.4; 1980: 5.1 (additive).

3. (a) 43.167, 41.333, 42.133, 40.5, 36.533, 36.467, 37.667, 40.5, 38.867; 42.58, 40.2, 39.42, 38.56, 38, 38.44, 37.58.

4. (a) 7.9, 9.5, 10.6, 6.4 (additive).

5. (a) 26.75, 28, 28, 28.375, 29.125, 29.875, 30.875, 31.625, 32.5, 33, 33.25, 34, 34.875, 35.375
   (b) 9.13, −15.17, −10.48, 16.52

6. 31, 30.2, 30.8, 32, ..., 39.6, 40.2.

7. (a) 1972: 50; 1979: 36.

8. 1.463, 0.806, 0.636, 1.090 (multiplicative).

9. (a) −11.7, 37.3, −25.6 (additive) or 0.89, 1.35, 0.75 (multiplicative)
   (b) 0.78 per month   (c) 63

10. (a)

Home	103.5	105.25	104.5	105	106.5	107	109	110.5	112
Export	101.75	104	105.75	107.25	109.75	112.25	116.5	121	123.5

11. Trend: 5.83, 5.981, 6.174, 6.38, 6.559, 6.751, 6.985, 7.233, 7.439, 7.64, 7.906, 8.195
    Seasonal variations: −1.263, −2.134, 3.146, 0.252.

12. (a) 400, 490, 540, 320 (additive).

13. (a) 30, 31, 35, 40, 45, 50, 54, 57
    (b) −18.25, 2.75, 29.75, −14.25   (c) £56 000, £56 000

14. (b) 21%, 45%, 25%, 13% (multiplicative).

15. (b) £114 000, £68 000, £172 000, £80 000 (additive).

16. (b) −32, −66, −104.67, 365; 0.897, 0.800, 0.701, 1.986; indices.

SECTIONS 7.1 TO 7.3

1. 116.4, 131.6, 144.4, 156.7.

2. (a) (i) 143.75, 175, 168.75, 200   (ii) 143.75, 121.74, 96.43, 118.52
   (b) (i) 105, 111.25, 107.5, 106.25, 110
       (ii) 105, 105.95, 96.63, 98.84, 103.53.

3. (a) 114.16 (or 110.96)   (b) 114.72 (or 108.86)

4. (a) (i) 121.88   (ii) 119.63   (b) (i) 111.49   (ii) 104.84.

5. (a) 110.56   (b) 109.99   (c) 110.28.

6. 105.74 (Paasche).

7. 97.78, 103.33.

8. 110.09, 110.47.

9. Prices: 93.55, 94.17; Quantities: 96.53, 97.17.

PAST QUESTIONS 7

1. (a) 108.1, 103.3   (b) 6.6.

2. 152.4, 160.5; increase of 315%.

3. 145.7.

4. 159.2.

5. (b) (i) 129.6   (ii) 115.0.

6. (a) 99.5   (b) 111.4   (c) 11% increase.

7. (b) 296.3.

8. 192.8, 190.2.

9. (a) 143.9, 192.9, 238.6, 304.2.

10. (b) 110.5, 122.6.

11. (b) 104.5.

12. (b) (i) 116.9   (ii) 128.4.

13. 128.2, 124.3.

14. (b) Prices: 155.65, 210.43; Quantities: 124.35, 94.78; Values: 193.55, 199.45.

15. (b) (i) 142.86, 197.14   (ii) 110, 124.67.

Summary of Formulae

CHAPTER 1

1.6 Arithmetic mean: $\bar{x} = \dfrac{\Sigma x}{n} = \dfrac{\Sigma fx}{\Sigma f}$

1.8 Median $= \frac{1}{2}(n+1)^{\text{th}}$ number

1.10 Lower quartile: $Q_1 = \left(\dfrac{n+1}{4}\right)^{\text{th}}$ number

Upper quartile: $Q_3 = \dfrac{3(n+1)^{\text{th}}}{4}$ number

Inter-quartile range: $\text{IQR} = Q_3 - Q_1$

Quartile deviation $= \frac{1}{2}(Q_3 - Q_1)$

1.11 Mean deviation $= \dfrac{\Sigma |x - \bar{x}|}{n} = \dfrac{\Sigma f |x - \bar{x}|}{\Sigma f}$

1.12 Standard deviation: $s = \sqrt{\dfrac{\Sigma (x - \bar{x})^2}{n}}$

$s = \sqrt{\dfrac{\Sigma f (x - \bar{x})^2}{\Sigma f}} = \sqrt{\dfrac{\Sigma fx^2}{\Sigma f} - \left(\dfrac{\Sigma fx}{\Sigma f}\right)^2}$

Variance $= s^2 = \dfrac{\Sigma fx^2}{\Sigma f} - \left(\dfrac{\Sigma fx}{\Sigma f}\right)^2$

CHAPTER 2

2.1 $p(X) = \dfrac{\text{number of ways X can occur}}{\text{total number of possible outcomes}}$

$p(\bar{X}) = 1 - p(X)$

2.2 $p(X \text{ and } Y) = p(X) \cdot p(Y)$

$p(X \text{ or } Y) = p(X) + p(Y)$

2.3 Factorials: $n! = n(n-1)(n-2) \ldots 3 \times 2 \times 1$

Combinations: $^nC_r = \dfrac{n!}{r!(n-r)!} = \begin{pmatrix} n \\ r \end{pmatrix}$

2.4 Binomial: $p(r) = {}^nC_r \cdot p^r \cdot (1-p)^{n-r}$

Mean $= np$, s.d. $= \sqrt{np(1-p)}$

2.5 Poisson: $p(r) = \dfrac{e^{-\lambda} \cdot \lambda^r}{r!}$

Mean $= \lambda$, s.d. $= \sqrt{\lambda}$

2.6 Normal: $u = \dfrac{x - \mu}{\sigma}$

95% confidence limits $= \mu \pm 1.96\sigma$

CHAPTER 3

3.1 $\mu_{\bar{x}} = \mu$, $\sigma_{\bar{x}} = \dfrac{\sigma}{\sqrt{n}}$

3.2 $u = \dfrac{\bar{x} - \mu_{\bar{x}}}{\sigma_{\bar{x}}} = \dfrac{\bar{x} - \mu}{\sigma/\sqrt{n}}$

3.3 t-test: $t = \dfrac{\bar{x} - \mu}{s/\sqrt{n-1}}$, $\nu = n - 1$

3.4 $t = \dfrac{|\bar{x}_1 - \bar{x}_2|}{\hat{\sigma}\sqrt{1/n_1 + 1/n_2}}$ where $\hat{\sigma}^2 = \dfrac{n_1 s_1^2 + n_2 s_2^2}{n_1 + n_2 - 2}$ with $\nu = n_1 + n_2 - 2$

3.5 $t = \dfrac{\bar{d}}{s_d/\sqrt{n-1}}$, $\nu = n - 1$

3.6 $\mu_{\hat{p}} = p$, $\sigma_{\hat{p}} = \sqrt{\dfrac{p(1-p)}{n}}$

$u = \dfrac{\hat{p} - \mu_{\hat{p}}}{\sigma_{\hat{p}}}$

$u = \dfrac{|\hat{p}_1 - \hat{p}_2|}{\sqrt{\hat{p}(1-\hat{p})(1/n_1 + 1/n_2)}}$ where $\hat{p} = \dfrac{n_1 \hat{p}_1 + n_2 \hat{p}_2}{n_1 + n_2}$

CHAPTER 4

4.1 $\chi^2 = \sum \dfrac{(O - E)^2}{E}$, $\nu = n - r$

4.2 In an $m \times n$ contingency table $\nu = (m-1)(n-1)$

4.3 When $\nu = 1$, $\chi^2 = \sum \dfrac{Y^2}{E}$ where $Y = |O-E| - \tfrac{1}{2}$

CHAPTER 5

5.2 PMCC: $r = \dfrac{\Sigma xy - n\bar{x}\bar{y}}{\sqrt{[\Sigma x^2 - n\bar{x}^2][\Sigma y^2 - n\bar{y}^2]}}$

5.3 Rank correlation coefficient: $r = 1 - \dfrac{6\Sigma d^2}{n(n^2-1)}$

5.4 $t = \dfrac{r\sqrt{n-2}}{\sqrt{1-r^2}}$, $\nu = n-2$

5.5 $y = a + bx$ where $b = \dfrac{\Sigma xy - n\bar{x}\bar{y}}{\Sigma x^2 - n\bar{x}^2}$ and $a = \bar{y} - b\bar{x}$

CHAPTER 6

6.3 Additive model: $X = T + S$ (or $T + S + R$)

6.4 Multiplicative model: $X = T \cdot S$ (or $T \cdot S \cdot R$)

6.5 Exponential smoothing: $T_i = \alpha X_i + (1-\alpha) T_{i-1}$

CHAPTER 7

7.1 Simple relative $= \dfrac{p_i}{p_0} \times 100$

Average relative: $= \dfrac{\Sigma(p_i/p_0)}{n} \times 100$

Weighted average $= \dfrac{\Sigma w(p_i/p_0)}{\Sigma w} \times 100$

7.2 Simple aggregate $= \dfrac{\Sigma p_i}{\Sigma p_0} \times 100$

Laspeyre: $\dfrac{\Sigma p_i q_0}{\Sigma p_0 q_0} \times 100$

Paasche: $\dfrac{\Sigma p_i q_i}{\Sigma p_0 q_i} \times 100$

Typical year: $\dfrac{\Sigma p_i q_t}{\Sigma p_0 q_t} \times 100$

Marshall-Edgeworth: $\dfrac{\Sigma p_i (q_0 + q_i)}{\Sigma p_0 (q_0 + q_i)} \times 100$

Fisher's ideal: $\sqrt{\left(\dfrac{\Sigma p_i q_0}{\Sigma p_0 q_0}\right)\left(\dfrac{\Sigma p_i q_i}{\Sigma p_0 q_i}\right)} \times 100$

7.3 Simple aggregate: $\dfrac{\Sigma q_i}{\Sigma q_0} \times 100$

Laspeyre: $\dfrac{\Sigma q_i p_0}{\Sigma q_0 p_0} \times 100$

Paasche: $\dfrac{\Sigma q_i p_i}{\Sigma q_0 p_i} \times 100$

INDEX

Additive model 116, 123
Adjusted frequency 3
Agreement 95, 105
Alternative 55, 58, 89
Areas
    in histogram 3
    under Normal curve 43
Arithmetic mean 12–13
Association 80, 83, 92, 102
Assumption 55, 58
Average
    deviations 118, 119, 120
    indices 125, 126
    relative 136
Averages 12–16

Bar charts 8–10
Base
    date 137
    price 136
    quantity 139
Basic probability 33
Best straight line 99, 116
Binomial approximation 40
Binomial distribution 37–9

Central value 18
Centred moving averages 114
Chain base 137, 145
Chi-squared (χ^2) distribution 74–86
Class intervals 1, 3
Coding 23–4
Combination of classes 77, 78, 82
Combination of events 33–5
Combinations 36–7
Combined indices 138, 146
Common variance 59
Component bar chart 9
Confidence limits 46, 54, 66, 74
Contingency tables 79–83
Corrected deviations 118, 119, 120
Correlation 91–104
Correlation coefficient 92–4
Critical region 89
Cumulative frequency curves 6–7
Current
    price 136
    quantity 140

Data 1
Degrees of freedom 57, 60, 74, 79, 97
Determination 106
Deviations 116, 117
Diagrams 2–10
Difference between means 59–62

Difference between proportions 66
Dispersion 17–24
Distribution of proportions 65–9
Distribution of sample means 53–4
Distributions
    binomial 37–9
    χ^2 74–86
    frequency 1
    Poisson 39–41
    Normal 42–6
    t 57–9

Estimate of variance 60
Estimation from ogive 7
Exclusive events 33
Expected frequency 74, 80
Exponential smoothing 127–9

Factorial 36
Fisher's ideal index 140, 142
Fixed base 137, 145
Fluctuations 132, 133
Forecasting 111–30
Frequency density 3, 6
Frequency distributions 1–25
Frequency polygons 5–6

Goodness of fit 74–8
Grouped frequency distribution 1

Heights of blocks 3
Histograms 2–4
Hypotheses: H_0, H_1 55, 56, 58

Independent events 33
Index numbers 136–47
Inter-quartile range (IQR) 18–20
Intervals 1
Inverse correlation 94

Laspeyre index 139, 141, 143
Level of significance 55
Linear
    correlation 93, 94, 97, 99
    relationship 92, 99
Lower quartile 18

Marshall-Edgeworth index 140, 142
Mean 12–13, 37, 39, 42
Mean deviation 20–1
Median 15–16
Modal class 14
Mode 13–14
Moving averages 112–16
Multiple bar chart 9, 10

Multiplicative model 122–7
Mutually exclusive events 33

Non-parametric tests 88
Normal distribution 42–6, 77
Null hypothesis 87, 89, 90

Observed frequency 74
Ogives 6–7
One-tailed test 56, 58

Paasche index 140, 141, 143
Paired t-test 62–4
Parametric statistics 88
Pie chart 8
Poisson distribution 39–41, 76
Polygons 5
Population standard deviation 60
Positive correlation 93, 96
Price aggregates 139–43
Price relatives 136–9
Probability 33–52
Product moment correlation coefficient (PMCC) 92–4
Proportions 65–9

Quantity indices 143–5
Quartile deviation 18
Quartiles 18

Random events 39
Range 17–18
Rank correlation 95–6
Regression 99–104, 116
Relationship 91, 96
Relatives 136–9
Residual
  index 122
  variation 116
Restrictions 74, 75, 77, 78

Sample
  means 53, 59
  size 53, 59

Sampling distributions 53–69
Scatter diagrams 91–2, 99
Seasonal index 122, 123, 124
Seasonal variations 116–21
Semi-inter-quartile range 18
Sigma (Σ) 12
Significance level 55, 74, 84, 89
Significance of the correlation coefficient 97–9
Significance of a sample mean 54–6, 58, 59
Simple relative 136
Smoothing constant 127
Spread 17–24
Standard deviation 22–3, 37, 39, 42
Standardised unit 43
Standard error 53

Tables 43
Tally chart 1
t-distribution 57–9, 60, 62
Test statistic 89
Time series 111, 112, 122, 123
Transformation 23
Trend 106, 111, 112
t-test 57, 60, 62, 97
Two-tailed test 55, 58
Typical year index 140, 143

Unequal class intervals 3
Unknown standard deviation 57
Upper quartile 18

Variance 22
Variance estimate 60
Volume indices 143–5

Weighted average 136, 139
Weighting 136, 139

$\bar{x}$ (arithmetic mean) 12–13

Yates' correction 83–4